Annals of Mathematics Studies
Number 56

ANNALS OF MATHEMATICS STUDIES

Edited by Robert C. Gunning, John C. Moore, and Marston Morse

1. Algebraic Theory of Numbers, *by* HERMANN WEYL
3. Consistency of the Continuum Hypothesis, *by* KURT GÖDEL
11. Introduction to Nonlinear Mechanics, *by* N. KRYLOFF *and* N. BOGOLIUBOFF
19. Fourier Transforms, *by* S. BOCHNER *and* K. CHANDRASEKHARAN
20. Contributions to the Theory of Nonlinear Oscillations, Vol. I, *edited by* S. LEFSCHETZ
21. Functional Operators, Vol. 1, *by* JOHN VON NEUMANN
24. Contributions to the Theory of Games, Vol. I, *edited by* H. W. KUHN *and* A. W. TUCKER
25. Contributions to Fourier Analysis, *edited by* A. ZYGMUND, W. TRANSUE, M. MORSE, A. P. CALDERON, *and* S. BOCHNER
28. Contributions to the Theory of Games, Vol. II, *edited by* H. W. KUHN *and* A. W. TUCKER
30. Contributions to the Theory of Riemann Surfaces, *edited by* L. AHLFORS *et al.*
33. Contributions to the Theory of Partial Differential Equations, *edited by* L. BERS, S. BOCHNER, *and* F. JOHN
34. Automata Studies, *edited by* C. E. SHANNON *and* J. MCCARTHY
38. Linear Inequalities and Related Systems, *edited by* H. W. KUHN *and* A. W. TUCKER
39. Contributions to the Theory of Games, Vol. III, *edited by* M. DRESHER, A. W. TUCKER *and* P. WOLFE
40. Contributions to the Theory of Games, Vol. IV, *edited by* R. DUNCAN LUCE *and* A. W. TUCKER
41. Contributions to the Theory of Nonlinear Oscillations, Vol. IV, *edited by* S. LEFSCHETZ
42. Lectures on Fourier Integrals, *by* S. BOCHNER. In preparation
43. Ramification Theoretic Methods in Algebraic Geometry, *by* S. ABHYANKAR
44. Stationary Processes and Prediction Theory, *by* H. FURSTENBERG
45. Contributions to the Theory of Nonlinear Oscillations, Vol. V, CESARI, LASALLE, *and* LEFSCHETZ
46. Seminar on Transformation Groups, *by* A. BOREL *et al.*
47. Theory of Formal Systems, *by* R. SMULLYAN
48. Lectures on Modular Forms, *by* R. C. GUNNING
49. Composition Methods in Homotopy Groups of Spheres, *by* H. TODA
50. Cohomology Operations, *lectures by* N. E. STEENROD, *written and revised by* D. B. A. EPSTEIN
51. Morse Theory, *by* J. W. MILNOR
52. Advances in Game Theory, *edited by* M. DRESHER, L. SHAPLEY, *and* A. W. TUCKER
53. Flows on Homogeneous Spaces, *by* L. AUSLANDER, L. GREEN, F. HAHN, *et al.*
54. Elementary Differential Topology, *by* J. R. MUNKRES
55. Degrees of Unsolvability, *by* G. E. SACKS
56. Knot Groups, *by* L. P. NEUWIRTH
57. The Atiyah-Singer Index Theorem: A Seminar, *by* R. S. PALAIS

KNOT GROUPS

BY

L. P. Neuwirth

PRINCETON, NEW JERSEY
PRINCETON UNIVERSITY PRESS
1965

L. C. Card 65-14393

Printed in the United States of America
by WESTVIEW PRESS, Boulder, Colorado

PRINCETON UNIVERSITY PRESS ON DEMAND EDITION, 1985

CONTENTS

CONTENTS

CHAPTER I

INTRODUCTION

§1. Introduction

The relations between topology and group theory have always been cordial at the very least, and this book is meant to take advantage of that fact. This work is intended to give some indication of the present state of knowledge concerning the fundamental group of the complement of an arbitrary polygonal knot in the three sphere. The relation of such a group to "group theory" on the one hand, and to the problems of three-dimensional manifolds on the other is an interesting one.

The homology groups of a knot group vanish in dimensions bigger then two, so that in a sense a knot group is close to being free. But, what is more to the point, the knot group arises in a natural way from a geometrical situation, and this situation permits the application of a good many geometric techniques. These techniques lead to algebraic results, and suggest algebraic techniques. So that, just as, classically, an abstract group theory arose from geometric considerations, one may hope that more sophisticated geometric attitudes will enrich group theory. Now recent results concerning three-dimensional manifolds have considerably advanced our knowledge , and the work of Wallace [67], and Lickorish [34], as well as the older work of Alexander [1], have placed knot theory in a more central location with respect to the theory of three-dimensional manifolds. I hope to present the reader some of the knot theory having to do particularly with the group of a knot. It will be seen that the geometry of the situation can rarely be ignored, and the interplay between algebra and geometry should be apparent.

Of fundamental importance to this presentation is the concept of a covering space. In order to present this idea coherently, I find it convenient to develop a theory of covering spaces of three-dimensional manifolds from a purely combinatorial point of view. This theory arises from an algorithm for computing a presentation of the fundamental group of the complement of a (possibly empty) one-dimensional subcomplex of a triangulated three-dimensional manifold. While a more general approach would have been possible, there seemed to be no need here to consider dimensions bigger than three.

One might adopt the view that a knot group is simply a peculiar special case of a group having a presentation of the following sort

$$(X_1, \ldots, X_n;\ X_{i_j}^{\epsilon_j} X_j X_{i_j}^{-\epsilon_j} = X_{j+1}) \quad j = 1, 2, \ldots, n-1, \quad \epsilon_j = \pm 1 \ .$$

This is shown for example in [54]. One may also derive the existence of such a presentation from results in Chapter III. This point of view may occasionally prove fruitful, but I have not been able to utilize it often enough in this book to make such a position worthwhile. It should not, however, be ignored as a possible entrée into the subject.

Material which has not appeared elsewhere includes all of Chapters VIII and III, and practically all unreferenced theorems.

Proofs are not given for all the theorems presented. Proofs are given when the methods of proof are interesting, or when the proofs have not appeared elsewhere. Occasionally a proof has been omitted because I do not understand it well enough to present it.

The reader familiar with knot theory will find at least the following omitted: Trotter's work on the cohomology of knot groups [63], the fascinating results of Crowell [8], Murasugi [39], and Kinoshita [31] on alternating knots, an exposition of a good deal of Crowell's recent investigations into the structure of the commutator module [9], Seifert's computations concerning the Alexander matrix [57], the relations between knots and braids, a discussion of the Stallings Fibrations [59], and finally links.

I thank Professors Fox, Papakyrikopoulos, Trotter, Stallings and Crowell for their generous help in the preparation of this manuscript.

The invaluable assistance of Peter Strom Goldstein is also gratefully acknowledged.

Finally, I wish to thank both Professor Gunning for his confidence and cooperation, and the Institute for Defense Analyses, whose atmosphere and encouragement provided a great stimulus for working on this material.

CHAPTER II

NOTATION AND CONVENTIONS

§1. Introduction

While most of the material in this chapter is a formality, there are some mildly unusual uses of familiar terminology. The careful reader is recommended to read §3 of this chapter before reading Chapter III.

§2. Group Theory

G will commonly denote a knot group and G' its commutator subgroup. However, $[H, H]$ and H' will be used interchangeably to denote the commutator subgroup of H.

Frequently reference will be made to a free product with amalgamation. We denote such a construction $A \underset{C}{*} B$, and depend upon the surrounding context to supply the information about the amalgamating maps $C \to A$, $C \to B$. The formal construction of $A \underset{C}{*} B$ may be found for example in [33]. There one also may find the main properties that will be used here.

Where confusion might arise we adopt the usual notation $(x_1, \ldots, x_n ; r_1, \ldots, r_m)$ and $|x_1, \ldots, x_n ; r_1, \ldots, r_m|$ to distinguish the presentation of a group from the symbol denoting the group itself.

The deficiency of a presentation is the number of generators minus the number of relations.

The cyclic group of order n is denoted Z_n.

The infinite cyclic group is denoted Z.

When a situation $1 \to F \to G \xrightarrow{\varphi} Z \to 0$ arises, (the extension, G, of the non-abelian group F, with identity 1, by the infinite cyclic group) we will frequently refer to a generator t of Z and behave as if $t \in G$. This may be done as long as care is exercised. The above sequence splits, so there exists a monomorphism, η from Z into G such that $\varphi\eta$ = identity. Thus by $t \in G$ we mean $\eta(t) \in G$. On the other hand η is not unique, for $\rho(t) = \eta(t) \cdot f$, for any $f \in F$, also determines a map (since Z is free) such that $\varphi\rho$ = identity.

In the same spirit, we may refer to the action of Z on F. By this we mean the cyclic group of automorphisms of F generated by conjugation of elements of F by $\eta(t)$ i.e., $f \to \eta(t) f\eta(t)^{-1}$. Here again, since η is not unique, this automorphism is only determined modulo an inner automorphism of F. (However, in many cases F will be abelian.) Occasionally we denote $v^{-1} av$ by a^v.

§3. Geometric Conventions

This book is concerned only with the fundamental group of the complement of <u>tame</u> (polygonal) knots in the 3-sphere. We will, especially in Chapter III, refer to combinatorial 3-manifolds.† By [38], any 3-manifold has a combinatorial structure and a small enough neighbourhood of a vertex in the interior of such a manifold is isomorphic to the cone over some triangulated 2-sphere.

Furthermore, in a combinatorial 3-manifold the star neighbourhood of an interior 1-simplex contains a cyclically ordered sequence of 3-simplices, as below.

---

† By a closed n-manifold is meant a compact separable Hausdorff space, such that every point has a neighbourhood homeomorphic to Euclidean n-space (E^n). An n-manifold with boundary is a compact separable Hausdorff space, such that every point has a neighbourhood whose closure is homeomorphic to the closed ball $\sum_{i=1}^{n} x_i^2 \leq 1$ in E^n.

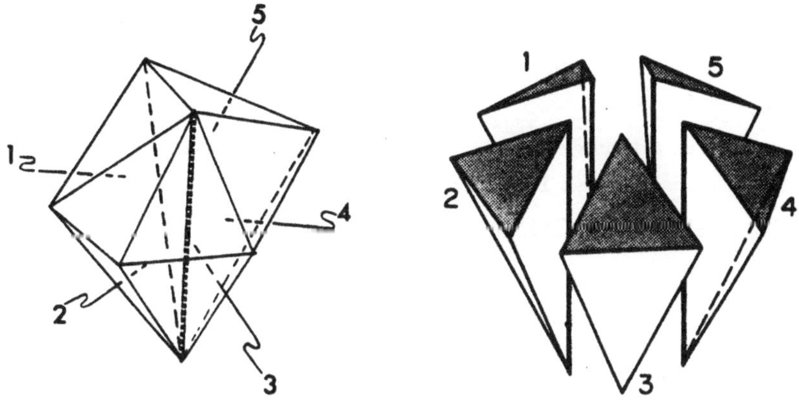

Fig. II-1

By the coboundary operator, δ, we understand the function from the set of i-simplices to subsets of the set of i+1-simplices, which associates to each i-simplex the set of i+1-simplices whose boundary contains the given i-simplex.†

Combining the notions of the last two paragraphs, we will, in Chapter III, define the ordered coboundary of a 1-simplex, given a small loop about that 1-simplex.

By a simplex we of course always mean a closed simplex. By an open simplex we mean a closed simplex minus its faces.

By a knot we mean a circle, imbedded as a 1-dimensional subcomplex in a 3-manifold which, unless otherwise stated, is meant to be the 3-sphere, S^3.

Two knots, k, k', are equivalent if there exists a homeomorphism of S^3 throwing k onto k'. This divides all knots into equivalence classes, and we refer to a class as a knot type.

We will sometimes consider a knot to have an orientation asso-

---

† This is the carrier of the usual mod 2 coboundary.

ciated with it. This situation will however be explicitly framed when it arises.

By a knot group we mean the fundamental group of the complement of a knot.

When a knot group is discussed it will be assumed, unless otherwise indicated, that it is _not_ the group of a trivial knot, that is, _the knot group is not infinite cyclic_.

CHAPTER III

COMBINATORIAL COVERING SPACE THEORY FOR 3-MANIFOLDS

§1. Introduction

A classical method for constructing <u>any</u> three-dimensional manifold consists in matching up 2-simplices in the boundary of a triangulated 3-cell. The proof of this fact depends on the construction of a "maximal cave" in the manifold.† The complement of this maximal cave (which must contain all the open 3-simplices) is the image of the boundary of the 3-cell under the identification mapping. Without referring directly to the construction of a three-dimensional manifold by this means, this chapter utilizes the complement of a maximal cave in a 3-manifold to compute the fundamental group of the manifold with a 1-dimensional subcomplex removed, and to construct (in a geometric way) covering spaces of this space.

§2. Computation of π_1 from a Maximal Cave

A well-known algorithm [58] for computing the fundamental group of a connected simplicial complex consists roughly in

1. constructing a maximal tree in the 1-skeleton;
2. generating a free group on the missing edges;
3. writing down one relation for each 2-simplex.

This process may be dualized if the simplicial complex is a manifold. In particular let us suppose we are given a triangulated, connected 3-manifold,

---

† It also depends upon the fact that a 3-manifold can be triangulated.

M. We shall compute $\pi_1(M)$ by this dual process. We begin by constructing a maximal cave in M. We select the interior of a 3-simplex, and break down the interior of the wall (i.e., the open 2-simplex) between that 3-simplex and an adjoining 3-simplex. We continue breaking down walls as long as we break into a 3-simplex we have not seen before. It is clear that we must eventually see all the 3-simplices, for our manifold is connected, and so the set of unseen 3-simplices always intersects the set of 3- simplices we have seen in some non-empty set of 2-simplices. Furthermore, it is clear by induction that our cave is simply connected, since each new 3-simplex is adjoined along an open 2-simplex wall.

Now, if we adjoin, one at a time, the interiors of the remaining 2-simplices, we obtain a space N which is nothing more than the complement of the 1-skeleton of M. Clearly $\pi_1(N)$ is free, and in fact contains the dual 1-skeleton of M as a deformation retract. (Formally one may deform the union of each open 3-simplex with its open 2-faces onto a 4-ad  in such a way that each open face is deformed into a point.) Not only is the fundamental group of this space free, but there is one free generator corresponding to each 2-face. These generators will later be described in more detail. Now we thicken slightly each open 1-simplex, and adjoin these thickened simplices to N one at a time. Application of the van Kampen theorem [66] gives, for each 1-simplex, one relation corresponding to a little loop around the 1-simplex. These will also be discussed in more detail. But before doing this we must adjoin a small neighborhood of each vertex to complete our computation. A small neighborhood of each vertex intersects the union of the open 3-simplices, 2-simplices and 1-simplices in a space homeomorphic to a 3-ball minus a point. This latter space being simply connected, no new relations are added when a neighborhood of each vertex is adjoined.

Now we may come back to our generators and relations and examine them more closely. The complement of our maximal cave is a 2-complex, which we denote by K. Recall that there is one generator corresponding to each 2-simplex in K. This generator is described by a path, beginning at a base point O in the interior of a 3-simplex, running up through the maximal cave to the 2-simplex in question, intersecting its interior once, and then running back to the base point along another path. The homotopy class of this loop is independent of the choice of path to and from the 2-simplex since the maximal cave is simply connected and a choice of ordering of the two 3-simplices in the coboundary of the 2-simplex fixes the direction of passage through the 2-simplex. So, by fixing an ordering of the coboundary of each 2-simplex in K, a particular set of generators is fixed. Now consider a 1-simplex. To each such there is a coboundary, and the set of 2-simplices in the coboundary of that 1-simplex is cyclically ordered by the order in which a small neat loop about the 1-simplex intersects the 2-simplices having that 1-simplex on their boundary. We call any such cyclically ordered coboundary a non-abelian coboundary. Since the maximal cave M - K is simply connected, the relation arising from the 1-simplex may be taken to be the word in the generators corresponding to

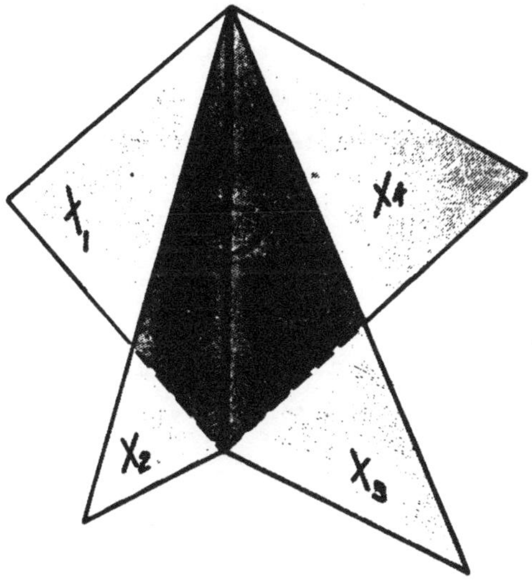

$X_1 X_2 X_3 X_4 = 1$

Fig. III-1

a non-abelian coboundary of the 1-simplex. (Those 2-simplices lying in the maximal cave correspond to the identity.) Since any two words arising from a cyclic ordering differ by a conjugation (i.e., $wa = a^{-1}(aw)a$) and possibly inversion, it does not matter which cyclic word we select as a relation. Figure III-1 illustrates the principle described above.

§3. The Splitting Complex

We may now immediately generalize the above described procedure in two directions.

1) By leaving out any set of closed 1-simplices, and not adjoining relations corresponding to these 1-simplices, we obtain the fundamental group of the complement of this set of 1-simplices.

2) Any closed 2-complex whose complement is connected and simply connected, may serve to give generators and relations from its 2-simplices and 1-simplices respectively. This is because the only properties of the maximal cave used were its connectedness, simple connectedness, and ownership of all the 3-simplices of M.

In view of this we make the following definition:

DEFINITION 3.3.1. Let L denote a closed 1-dimensional subcomplex of a connected 3-manifold M. A 2-dimensional subcomplex K of M is called a splitting complex† for (M, L) if

a) $M - K$ is connected;
b) $M - K$ is simply connected;
c) $K \supset L$.

We summarize our preceding discussion by means of the following theorem.

† In case L is empty K has also been called a spine of M. [72]

THEOREM 3.3.1. If K is a splitting complex for (M, L) then $\pi_1(M-L) \approx |x_1, \ldots, x_m; r_1, \ldots, r_n|$ where the x_i are in 1-1 correspondence with the 2-simplices of K and one ordering of the coboundary of each, and the r_i are in 1-1 correspondence with the 1-simplices of $K - L$. Furthermore, r_i is a word in the x_i which represents a non-abelian coboundary of the 1-simplex corresponding to that r_i.

The removal of a 1-dimensional subcomplex from M certainly does not affect the construction of a maximal cave (which is a union of simplices of dimension 2 and 3) so that we may state:

THEOREM 3.3.2. Given a 1-dimensional subcomplex $L \subset M$, there exists a splitting complex for (M, L).

§4. A Splitting Complex for a Knot

As an application of this algorithm it is appropriate to show how it may be used to compute a presentation for the fundamental group of the complement of a knot k in S^3. We do this by constructing a splitting complex for (S^3, k).

Projecting a knot k in general position on a 2-sphere $S^2 \subset S^3$ produces a "checker board" on the sphere. That is, we may color one set of regions into which S^2 is divided black, and the complementary set white, in such a way that any two regions with a common boundary are colored differently. Without loss of generality we may assume k lies on S^2 except in the neighborhood of double points. At each double point we assume one arc lies on one side of the S^2, and the other arc on the other side. Consider one such double point. Denote by a, b the end points of one arc ab, and by c, d the endpoints of the other cd. Assume $a, b, c, d, \in N$ where N is a nice neighborhood of the double point. Suppose a, c lie on the boundary of the same black region in N. Then b, d also lie on the boundary of a black region in N. We join a, c and b, d respectively by simple arcs ac, bd lying in these black regions. Then ac, cd, db, ba span in S^3 a ribbon with a half twist joining up two black regions.

If we substitute this ribbon for its projection on S^2 and make the same construction at every double point, then all the black regions may be hooked up by ribbons, and a surface spanned in the knot. This is one of two checkerboard surfaces. The other is constructed in the same manner using the white regions. These two surfaces may be constructed so that their intersection is the knot plus one simple arc joining each pair of double points. The diagram below describes the union of these two surfaces in the neighborhood of a double point.

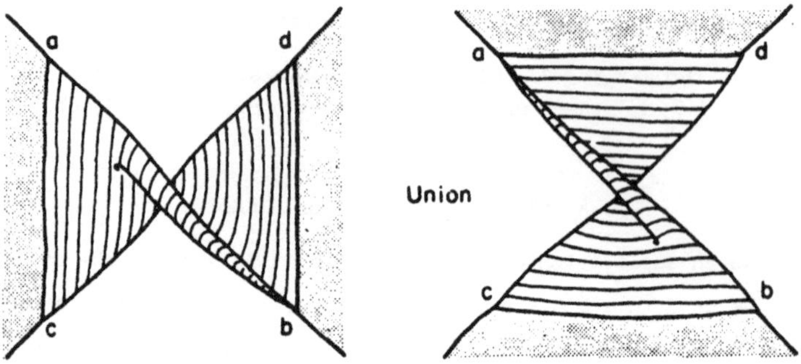

Fig. III-2

Now remove a disc from the union of these two surfaces. The resulting complex is easily seen to be a splitting complex for (S^3, k). Application of the algorithm described previously, and a little thought reveals that each region gives rise to a generator, and each double point to a relation. This presentation is the classical Dehn presentation. A sample relation is shown below.

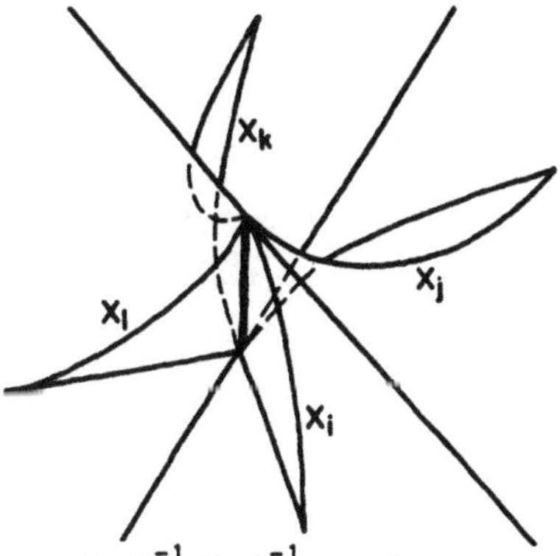

Each X in this picture corresponds to a 2-simplex and ordering of coboundary top to bottom.

$x_\ell x_k^{-1} x_j x_i^{-1} = 1$

Fig. III-3

As splitting complex we might also have chosen the join of a point with the knot. In which case we would obtain the classical Wirtinger presentation.

Calculation of the Euler characteristic of the decomposition of a sphere effected by the projection of a knot on the sphere implies that the <u>deficiency of the presentation we have obtained is</u> +1.

§5. Construction of Coverings from a Splitting Complex.

We now utilize the splitting complex to construct coverings of M - L.

Recall the classical definition of a covering space $\tilde{X}$ of X.

1. There is a continuous map p, of $\tilde{X}$ onto X.
2. Given a point $S \in X$, there exists a neighborhood, U, of S, and a neighborhood U_σ of each point σ in $p^{-1}(S)$, such that
   a) $p|U_\sigma$ is a homeomorphism onto U.
   b) The U_σ are disjoint.

We may summarize (2) by saying each point in X has a neighborhood which is evenly covered.

Expressed in more general terms $\tilde{X}$ is a locally trivial fiber bundle over X, with discrete fiber.

K, a splitting complex of (M, L) will be used to construct combinatorial coverings of (M-L). A combinatorial covering is a covering in which all the spaces and mappings are simplicial.

In order to construct a covering of M - L we need first to split M open along K. This process is analogous to the familiar process of cutting a 2-manifold along a set of simple curves to obtain a disc. We proceed as follows:

Let M^1 denote a union of disjoint (closed) 3-simplices in 1-1 correspondence with the closed 3-simplices of M. We make M split along K, by matching two 3-simplices of M^1 along a (closed) face if the corresponding 3-simplices of M are incident along the corresponding face, and if that face is not in K. This process leads to a complex $\hat{M}$, and a natural map φ: $\hat{M} \to M$ which matches up a pair of (closed) 2-simplices of $\hat{M}$ corresponding to a 2-simplex in K. (There are pairs of simplices matched because M is a manifold, and each 2-simplex is incident to precisely two 3-simplices.)

Having split M along K, we may now construct a covering of M - L. The data required for the construction of such a covering is an assignment of a permutation (possibly of infinite degree) to each 2-simplex of K, and ordered pair of 3-simplices of M incident along that 2-simplex. This assignment must satisfy the condition that the product of the permutations corresponding to a non-abelian coboundary of a 1-simplex in K - L shall be the identity permutation. (Of course if we assign a permutation π to a 2-simplex, and an ordering of its coboundary, then we assign π^{-1} to that 2-simplex along with the opposite ordering of its coboundary.) Having suggested informally what we wish, let us proceed formally.

We now have at our disposal M split along K, which we denoted by $\hat{M}$, and a map φ: $\hat{M} \to M$.

If α denotes an interior 2-simplex in M then we define $\bar{\delta}(\alpha) = (\beta, \gamma)$ and $-\bar{\delta}(\alpha) = (\gamma, \beta)$ where β and γ are the distinct 3-simplices in M having α as a face. Thus with each 2-simplex α, we associate some ordering of the coboundary of α. (This of course is equivalent to orienting each interior 2-simplex in M.)

The previously discussed ordered (or non-abelian) coboundary of a 1-simplex, τ, we denote also by $\bar{\delta}(\tau)$. $\bar{\delta}(\tau)$ is to be any representative coboundary.

Now suppose we are given a set A (possibly infinite) and a function, P, from the set of all pairs $(\alpha, \bar{\delta}(\alpha))$, where α is a 2-simplex of K, to the group of all permutations of A. We extend P to pairs $(\alpha, -\bar{\delta}(\alpha))$ by the condition $P(\alpha, -\bar{\delta}(\alpha)) = [P(\alpha, \bar{\delta}\alpha)]^{-1}$.

If $\bar{\delta}(\tau) = (\alpha_1, \ldots, \alpha_n)$ for a 1-simplex τ in K-L, and a small loop about τ passes through α_i from the 3-simplex β_{2i-1} to the 3-simplex β_{2i}, then the informally stated requirement mentioned earlier is now precisely stated as

$$P(\alpha_1, (\beta_1, \beta_2))\, P(\alpha_2, (\beta_3, \beta_4)) \ldots P(\alpha_n, (\beta_{2n-1}, \beta_1)) = 1 .$$

If this condition is satisfied for one representative coboundary it clearly is satisfied for any coboundary. We now proceed to construct a candidate for a covering space. Let $\hat{M}_A$ denote the disjoint union of copies of $\hat{M}$ indexed by the elements of A. We shall occasionally refer to these copies as sheets. Denote by φ_a the map φ defined on the copy $\hat{M}_a$. Now each copy, $\hat{M}_a$ of $\hat{M}$ has two 2-simplices corresponding to each 2-simplex in K. We identify in the natural way a (closed) 2-simplex, σ, of $\hat{M}_a$ with a (closed) 2-simplex, σ_b of $\hat{M}_b$ when and only when

1. $\varphi_a(\sigma_a)$, $\varphi_b(\sigma_b)$ are the same 2-simplex, σ in K.
2. $\bar{\delta}(\sigma) = ((\varphi_a(\bar{\delta}(\sigma_a)), \varphi_b(\bar{\delta}(\sigma_b)))$.
3. $P(\sigma, (\varphi_a(\bar{\delta}(\sigma_a)), \varphi_b(\bar{\delta}(\sigma_b)))$ maps a to b .

This construction leads to a rule telling you which copy of $\hat{M}$ you walk into when you are in a copy of $\hat{M}$ and you cross a 2-simplex corresponding to a 2-simplex of K. This means we may associate to a path in $\hat{M}$, and a sheet $\hat{M}_a$, a path in our identification space, by walking along a copying path which starts in $\hat{M}_a$ and enters a different sheet when our assignment says to (recalling the process of analytic continuation).

Now remove from this identification space all 1-simplices which are mapped onto L by the various φ_a.

Having performed all the identifications and removals we were supposed to, we denote by $\tilde{M}$ the resulting space, and by Φ the mapping from $\tilde{M}$ to $M - L$ induced by the mappings φ_a. In what follows we may consider the open 3-simplices of $\hat{M}_a$ to be contained in $\tilde{M}$.

THEOREM 3.5.1. $(\tilde{M}, \Phi)$ is a covering space of M.

PROOF. Obviously, Φ is continuous and maps $\tilde{M}$ onto M.

1. If S is a point in the interior of a 3-simplex of $M - L$, then the interior of that 3-simplex is evenly covered by a copy of that 3-simplex in each copy of $\hat{M}$.

2. If S is a point in the interior of a 2-simplex, E, then the (open) coboundary of that (open) 2-simplex is a neighborhood of S that is evenly covered. For if E is not in K then each copy of $\hat{M}$ owns a copy of δE, while if E is in K either two distinct copies of $\hat{M}$ are incident along a 2-simplex E' mapped by Φ into E, or a single copy of $\hat{M}$ owns a copy of δE, depending upon the action of the permutation assigned to E, on the index of the copy of $\hat{M}$.

3. If S is a point in the interior of a 1-simplex T, not contained in L, then the set of all (open) 3-simplices in $M - L$ having T on their boundary is a neighborhood of S which is evenly covered. This may be shown as follows, if we run around T along a small oriented loop L beginning in a 3-simplex Γ_1 and not intersecting any 1-simplices, we pass successively through 3-simplices, $\Gamma_1, \Gamma_2, \ldots, \Gamma_n, \Gamma_1$. If we select an index $\alpha \in A$ then we may denote by $\alpha_1, \alpha_2, \ldots, \alpha_n, \alpha_{n+1}$ the successive images of α_i induced by the permutations corresponding to the 2-simplices in $\delta(T)$ lying in K, and the ordering of the coboundary of these 2-simplices induced by our little oriented loop. By our condition on permutation assignments $\alpha_{n+1} = \alpha$. Now for each selection of an index $\beta_1 \in A$, fixing L there is induced in this way a unique sequence $\beta_1, \beta_2, \ldots, \beta_n$. Furthermore, if $\alpha_1 \neq \beta_1$, $\alpha_j \neq \beta_j$. It follows from this that the interior of $\bigcup_{i=1} \Gamma_i$ is evenly covered by the interior of the disjoint sets of the form $\Gamma_1^{\beta_1} \cup \Gamma_2^{\beta_2} \cup \cdots \cup \Gamma_n^{\beta_n}$, where $\Gamma_i^{\beta_i}$ is the unique

3-simplex in $\hat{M}_{\beta_1}$ lying over Γ_1.

4. Suppose now that s is a vertex in $M - L$. As $M - L$ is a manifold, the star of s is a triangulated 3-ball. Select a smaller 3-ball B (containing s) whose boundary is a triangulated 2-sphere which we denoted by H, the triangulation being induced by the 2-skeleton of M. Name the 3-simplices having s as a vertex $\Lambda_1, \Lambda_2, \ldots, \Lambda_m$. Select a 3-simplex $\Lambda_1^{\alpha_1}$ in $\hat{M}_{\alpha_1}$ lying over Λ_1.

The interior of any other 3-simplex, Λ_r, in the star of s may be reached by a path on H starting in Λ_1 and intersecting only 2-simplices in H. Such a path determines a sequence of indices, $\alpha_1, \alpha_{1_2}, \ldots, \alpha_{1_\ell}$, as it passes from one 3-simplex to another through 2-simplices. These indices are the sequence of images of α_1 determined by our permutation assignment. Now having selected $\Lambda_1^{\alpha_1}$ over Λ_1, we claim that the determination of $\Lambda_r^{\alpha_{1_\ell}}$ over Λ_r is independent of the path chosen from Λ_1 to Λ_r. This is true because two paths α, and β from Λ_1 to Λ_r determine a loop $L = \alpha\beta^{-1}$ and any loop L on H not intersecting the 0-skeleton of H may be deformed in the complement of the 0-skeleton of H, into a product of loops encircling each vertex of H once, and each of these loops induces the identity permutation on A. This shows that the selection of an index α_1 and a 3-simplex Λ_1 determines uniquely a neighborhood of a preimage of s. By our construction, any identification of vertices arises from an identification of some pair of 2-simplices having these vertices on their boundary, so that the neighborhoods determined by distinct α_1 and fixed Λ_1 are disjoint and the interior of the star of s is evenly covered by a set of neighborhoods which may be indexed by the elements of A.

This completes the proof of the theorem.

At this point a splitting complex for (M, L) has been utilized for two ends; first to obtain a presentation of $\pi_1(M-L)$, and secondly to obtain certain coverings of $M - L$. We will unify these two uses by means of the following theorem, which we must prove to show the utility of this combinatorial theory.

THEOREM 3.5.2. [Fundamental theorem of Covering Space Theory; (for combinatorial 3-manifolds with a 1-dimensional subcomplex removed).] To every subgroup, $S \subset \pi_1((M-L), o)$ there corresponds a combinatorial covering space $(\tilde{M}, \Phi)$, constructed from any splitting complex and such that $\Phi^*(\pi_1(\tilde{M}, \tilde{o})) = S$; conversely, to each connected combinatorial covering, $(\tilde{M}, \Phi)$ there corresponds a subgroup $S \subset \pi_1(M-L)$ such that $\Phi^*(\pi_1(\tilde{M})) = S$.

PROOF. A combinatorial covering is a classical covering by Theorem 3.5.1 so the second part of the theorem is true.

Let K be a splitting complex for (M, L).

Suppose $S \subset \pi_1((M-L), o)$ is given. Assume o lies in the interior of a 3-simplex. Let A denote the set of left cosets of S. Now according to Theorem 3.3.1 to each 2-simplex E of K and ordered pair of 3-simplices in $\bar{\delta}(E)$ there corresponds a generator of $\pi_1(M-L)$. On the other hand, to each generator of $\pi_1(M-L)$ there corresponds a permutation of the cosets S_g of S, determined by right multiplication of a coset by that generator. These two correspondences give rise to an assignment of permutations of A to the 2-simplices of K and the ordered coboundary of each. The permutation induced by a small closed loop about a 1-simplex in $M - L$ is the identity because the product of the corresponding generators is the identity of $\pi_1(M-L)$ and hence leaves the cosets S_g of S fixed.

Thus by Theorem 3.5.1 we may construct a covering from the permutation assignment we have made. Denote by $(\tilde{M}, \Phi)$ this covering. Let us use the same notation as before for the copies of M split along K, that is, $\hat{M}_\alpha$ $\alpha \in A$.

Take as base point in $\tilde{M}$ the unique point $\tilde{o}$ lying over o and located in $\hat{M}_S$ (the copy of M split along K with index the coset of S which is S itself). We need to prove $\pi_1(\tilde{M}, \tilde{o})$ is mapped by Φ^* onto S.

Any closed curve, $\widetilde{c}$, based at $\widetilde{O}$ is mapped by Φ into a closed curve c based at O, and furthermore, the permutation induced by this latter curve (as in the construction preceeding the proof of Theorem 3.5.1, a curve in M induces a permutation of A) must leave the coset S fixed since we find ourselves back in $\hat{M}_s$ when starting there and tracing the curve $\widetilde{c}$ over c. But if the permutation induced by c leaves S fixed, then of course c represents an element of S. Thus, $\Phi^*\left(\pi_1(\widetilde{M}, \widetilde{O})\right) \subset S$.

Conversely if c is a curve representing an element of S, then the permutation it induces must of course leave the cosets of S fixed, so that c may be lifted to a closed curve $\widetilde{c}$ based at $\widetilde{O}$. Thus, $\Phi^*\left(\pi_1(\widetilde{M}, \widetilde{O})\right) \supset S$.

This completes the proof of the theorem.

Before making an amusing application of this means of constructing coverings we make the following observation which may be easily verified by the reader.

> If K is a 2-complex in a 3-manifold M, and M - K is connected, then an assignment of a permutation to each 2-simplex of K and ordering of its coboundary, satisfying the condition that the product of permutations corresponding to the non-abelian coboundary of a 1-simplex in K - L is the identity, may be used to construct a covering of M - L. In other words, if we weaken the conditions for a splitting complex so that M - K is not required to be simply connected and K is not required to contain L, we still may construct coverings of M - L from usable permutation assignments.

The reason M - K was required to be simply connected was so that all subgroups of π_1(M-L) could be used to construct coverings, if however we are willing to accept something less than that, then M - K needn't be simply connected.

Now, suppose a 2-manifold, T, is imbedded semi-linearly in a 3-manifold M in such a way that M - T is connected. Then assigning the permutation (12) to each 2-simplex of T will permit the construction of a 2-sheeted covering of M since $(12)^{-1} = (12)$, and each coboundary of a 1-simplex in T contains precisely two 2-simplices. Thus we have proved

THEOREM 5.5.3. If a 2-manifold can be semi-linearly imbedded in a 3-manifold, M, without separating M, then M has a connected two-sheeted covering.

COROLLARY 5.5.4. Any 2-manifold semi-linearly imbedded in L(2n+1, q) (lens space) separates.

PROOF. $\pi_1\big(L(2n+1, q)\big) \approx Z_{2n+1}$ which has no subgroups of index 2, so L(2n+1, q) has no connected two-sheeted cover.

As a final application of splitting complexes, an irregular covering of a knot is described by means of the picture below; there are three sheets and the permutations assigned to a 2-simplex and its ordered coboundary "top to bottom" are as indicated. These assignments determine all the other assignments to 2-simplices and their ordered coboundaries.

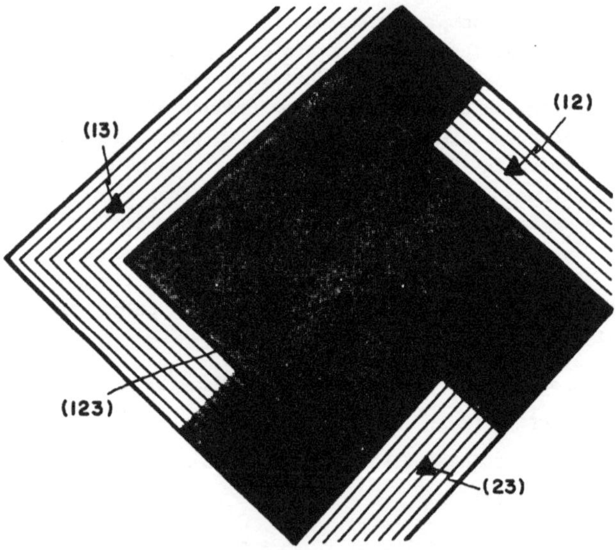

Fig. III-4

Before passing on to some applications of the ideas in this chapter, we may define

A. The Monodromy Group of a covering corresponding to a subgroup $S \subset \pi_1$ is the group of permutations of the cosets S_g of S in π_1 effected by right multiplication by elements of π_1; it is also the representation of π_1 as a group of permutations of the sheets determined by mapping an element g of π_1 onto the permutation of the sheets determined by walking in M along a closed curve representing g.' The equivalence of these two definitions follows from the proof of Theorem 5.5.1.

B. The Group of Covering Translations of a covering is the group of permutations of the left cosets of S in its normalizer, $N(S)$, induced by left multiplication by elements of $N(S)$ of these left cosets of S in $N(S)$. It is also the group of those permutations of the sheets $\hat{M}_\alpha$ which define a homeomorphism of $\tilde{M}$.

Reiterating, the group of covering translations is on the one hand the normalizer of S modulo S, and on the other the group of all permutations of the sheets which define homeomorphisms of $\tilde{M}$. This follows from the fact that a permutation of the sheets of $\tilde{M}$ defining a homeomorphism of $\tilde{M}$ is completely determined when the image of $\hat{M}_S$ is known; for if $\hat{M}_S$ is mapped onto $\hat{M}_{Sg}$ then $\hat{M}_{Sh}$ is mapped onto $\hat{M}_{Sgh}$ where h is a generator of $\pi_1(M)$. This implies $\hat{M}_{Sw}$ is mapped onto $\hat{M}_{Sgw}$ for any $w \in \pi_1$, so that the permutation defining the homeomorphism is induced by left multiplication of the left cosets of S, by elements g, which when multiplying any left coset of S on the left, give another left coset; but these g's are just the members of the normalizer of S. Since multiplication of S by an element g leaves the coset S fixed when and only when $g \in S$, the equivalence of the two definitions of the group of covering translations follows.

CHAPTER IV

THE COMMUTATOR SUBGROUP AND THE ALEXANDER MATRIX

§1. Introduction

In this chapter we shall first study the covering of the complement of a knot corresponding to the commutator subgroup of its group. This will lead in a natural way to the Alexander matrix.

Alexander's duality theorem implies that a knot group, G, modulo its commutator subgroup is infinite cyclic. As remarked in Chapter III, the presentation for G we have obtained has deficiency +1. This state of affairs simplifies the study of the Alexander Matrix of G. We shall discuss this matrix, as well as its elementary divisors, and it will be seen that the matrix determines G modulo its second commutator subgroup. The determinant of this matrix is the Alexander Polynomial and it will also be examined. We begin however, by proving a theorem which reveals the structure of the commutator subgroup of a knot group. Later in the chapter, Crowell's work on the abelianized commutator subgroup will be described.

§2. An Orientable Surface Spanned by a Knot

Let $\bar{M}$ denote the closure of the complement of a regular neighborhood† of a knot $k \subset S^3$. Then $\bar{M}$ is a manifold with boundary, and by Alexander Duality there exists an isomorphism φ from $H_1(\bar{M})$ onto Z.

† In the sense of [68].

The torus $\partial\bar{M}$ may be projected by a map ψ onto one of its factors S^1 in such a way that the following diagram is commutative.

$$\begin{array}{ccccc} H_1(\partial\bar{M}) & \xrightarrow{\text{inclusion}} & H_1(\bar{M}) & \longrightarrow & 0 \\ & \searrow^{\psi^*} & \downarrow \varphi & & \\ & & H_1(S^1) \approx Z & & \end{array}$$

By elementary obstruction theory ψ may be extended to $\bar{M}$. There is no loss of generality in assuming ψ is simplicial, so that by a well known principle [73] the preimage of an interior point p, of a 1-simplex of S^1, is a (possibly disconnected) 2-manifold whose boundary is contained in $\partial\bar{M}$. Since p locally disconnects S^1 it follows that the 2-manifold preimage of p is orientable. Discarding all components in the preimage of p but the one having a boundary, denote the one selected by the symbol N. Surely ∂N has only one component since $\psi^{-1}(p) \cap \partial(\bar{M})$ is a simple closed curve. Since this curve is simple and homologous to 0 in $\bar{M}$, it must be homologous to k in the closure of $S^3 - \bar{M}$ [69]. It follows that N may be stretched through $S^3 - \bar{M}$ to k. But more accurately, there exists a non-singular annulus A contained in the closure of $S^3 - \bar{M}$ such that $\partial A = k \cup \partial N$. Thus $A \cup N$ is a 2-manifold with boundary k.

More constructive and instructive means of finding an orientable surface spanned by k may be found in [57] and [18].

§3. The Infinite Cyclic Covering of a Knot

Let S denote an orientable surface contained in S^3 such that $\partial S = k$. Suppose that the genus of S is g, and that any other surface spanning k has genus greater than or equal to g. g <u>is called the genus of</u> k. Obviously g is a knot type invariant.

As S is orientable we may orient each 2-simplex in S and assign an ordering to the pair of 3-simplices in each coboundary of a 2-simplex in S, so that the induced orientation of the coboundary of these simplices agrees with some pre-assigned orientation of S^3. If σ

is a 2-simplex, denote this ordering $\bar{\delta}(\sigma)$. Now, assign the permutation of the integers $(n \to n+1)$ to each pair $(\sigma, \delta(\sigma))$, then this assignment defines, via the construction in Chapter III, a covering $\widetilde{M}$ of $S^3 - k$.

Any element $\alpha \in \pi_1(S^3 - k)$ having linking number [58] 0 with k belongs to the subgroup of this covering, for it will have intersection number [58] 0 with S, and this implies that the permutation of the integers (sheets) induced by α is trivial. Conversely, if α belongs to the subgroup of this covering, it follows that α must have linking number 0 with k. Thus we have constructed the covering of $S^3 - k$ corresponding to the commutator subgroup of $\pi_1(S^3 - k)$.

Particularizing the description in Chapter III, we obtain $\widetilde{M}$ by taking a countable number of disjoint copies of $X = (S^3$ split along $S)$, indexed by the integers, and pasting copy X_i to copy X_{i+1} along the "right hand" copy of S, ${}_iS_1$ in X_i and the "left hand" copy ${}_{i+1}S_2$, of S in copy X_{i+1}. The common copy of k in each X_i is then removed.

Copies of S^3 split along S

X_{i-1}	X_i		X_{i+1}
${}_{i-1}S_1$	${}_iS_2$	${}_iS_1$	${}_{i+1}S_2$

We shall use the notation of the above picture in the next section.

The group of covering translations of $\widetilde{M}$ is infinite cyclic since the commutator subgroup is normal. The group of covering translations may be generated by the permutation of the sheets sending X_i to X_{i+1}. Denote the covering translation sending X_i to X_{i+n} by t^n.

§4. A Property of the Surface of Minimal Genus

$(S^3$ split along $S) = X$ has on its boundary two copies of S, which we denote S_1, S_2, noting that $S_1 \cup S_2 = \partial X$, $S_1 \cap S_2 = k$.

LEMMA 4.4.1. The inclusion map $h: S_1 \to \partial X$ induces a monomorphism $h^{\#}: \pi_1(S_1) \to \pi_1(\partial X)$.

PROOF. Since k is knotted, S_1 is of genus greater than zero, hence no power of k considered as an element of $\pi_1(S_1)$ is null homotopic. Since $\partial X = S_1 \cup S_2$, and $S_1 \cap S_2 = k$, a simple application of the van Kampen Theorem gives

$$\pi_1(\partial X) \approx \pi_1(S_1) \underset{\pi_1(k)}{*} \pi_1(S_2) \quad .$$

Thus, $\pi_1(S_1)$ is imbedded monomorphically in $\pi_1(\partial X)$ by a map induced by the inclusion h.

LEMMA 4.4.2. The inclusion map $\nu: S_1 \to X$ induces a monomorphism $\nu^{\#}: \pi_1(S_1) \to \pi_1(X)$.

PROOF. Suppose the lemma is false, and that α is a closed curve on S_1, such that $\alpha \cong 0$ in X, $\alpha \not\cong 0$ on S_1.[†] According to Lemma 4.4.1. $\alpha \not\cong 0$ on ∂X. Since S_1 is polyhedral, α may be assumed polygonal so that X and α satisfy the hypothesis of the Loop Theorem ([50], Theorem 15.1 and Theorem 1), thus we may assume that α is simple. According to Dehn's Lemma [51], α bounds a non-singular polyhedral disc in X. If we cut S_1 along α, and sew discs to both sides of the cut we obtain a new surface S_1' which is bounded by k. If the curve α separates S_1, then because $\alpha \not\cong 0$ on S_1, the new surface S_1' has lower genus than S_1, which contradicts the assumption that the genus of S_1 is minimal. If α does not separate S_1, then compare the Euler Characteristic of S_1, $\chi(S_1)$, with that of S_1'. Since the cut adds one vertex, one edge, and two faces, $\chi(S_1') - \chi(S_1) = 2$, hence S_1' has lower genus than S_1, so we again arrive at a contradiction. The existence of the curve α thus leads to a contradiction, so the lemma is proved.

REMARK. Lemmas 1 and 2 obviously remain valid if S_2 is substituted for S_1. Lemma 2 also remains valid if k is removed from X.

---

† $\cong$ denotes homotopic to."

Roughly speaking, what this lemma means is that non-contractible curves on a surface S, of minimal genus spanning k are non-contractible in $S^3 - S$ if they are pushed off the surface (to either side).

§5. The Structure of the Commutator Subgroup of a Knot Group.

For the rest of this chapter G will denote $\pi_1(S^3 - k)$.

THEOREM 4.5.1. If [G, G] is finitely generated, it is free of rank 2g, where g is the genus of k.

If [G, G] is not finitely generated, then either it is:

A) a non-trivial free product with amalgamation on a free group of rank 2g, and may be written in the form

$$\cdots \underset{F_{2g}}{*} A \underset{F_{2g}}{*} A \underset{F_{2g}}{*} A \underset{F_{2g}}{*} A \underset{F_{2g}}{*} A \underset{F_{2g}}{\cdots},$$

where F_{2g} is free of rank 2g, and the amalgamations are all proper and identical, or

B) locally free, and a direct limit of free groups of rank 2g.

REMARK. I do not know if case B) can occur.†

PROOF. By virtue of Lemma 4.2, the last remark in §4 and the identification of ${}_iS_1$ and ${}_{i+1}S_2$, the following diagram is valid for every i,

$$\pi_1(X_i) \xleftarrow[{}_if_1]{} \pi_1({}_iS_1) \xrightarrow[{}_{i+1}f_2]{} \pi_1(X_{i+1})$$

where the ${}_if_j$ are monomorphisms.

† R. Crowell has informed me that E. Brown and he have proved that case B) cannot occur.

By a simple application of the van Kampen Theorem, the fundamental group of $X_i \cup X_{i+1}$ is the direct limit of the above system. This direct limit is a free product with amalgamated subgroup, $\pi_1(X_i) \underset{\pi_1({}_iS_1)}{*} \pi_1(X_{i+1})$. Let

$$Y_n = X_0 \cup X_1 \cup \cdots X_{n-1} \cup X_n, \qquad n \geq 0$$

$$Y_{-n} = X_{-1} \cup X_{-2} \cup \cdots X_{n+1} \cup X_{-n}, \qquad n \geq 1$$

$$Y_\infty = \bigcup_{i=1}^{\infty} X_i \qquad Y_{-\infty} = \bigcup_{i=-1}^{\infty} X_i \, .$$

Using the fact that each factor in a free product with amalgamation is contained as a subgroup in the free product with amalgamation [33] and proceeding inductively it is clear that from the above diagram one obtains,

$$(1) \quad \pi_1(Y_n) = \pi_1(Y_{n-1}) \underset{\pi_1(S)}{*} \pi_1(X_n)$$

$$= (((\pi_1(X_0) \underset{\pi_1(S)}{*} \pi_1(X_1)) \underset{\pi_1(S)}{*} \pi_1(X_2)) \cdots \underset{\pi_1(S)}{*} \pi_1(X_{n-1})) \underset{\pi_1(S)}{*} \pi_1(X_n) \, .$$

$$(2) \quad \pi_1(Y_{-n}) = \pi_1(Y_{-n+1}) \underset{\pi_1(S)}{*} \pi_1(X_{-n})$$

$$= (((\pi_1(X_{-1}) \underset{\pi_1(S)}{*} \pi_1(X_{-2})) \underset{\pi_1(S)}{*} \pi_1(X_{-3})) \cdots \underset{\pi_1(S)}{*} \pi_1(X_{-n+1})) \underset{\pi_1(S)}{*} \pi_1(X_{-n})$$

$\pi_1(Y_\infty) = \underset{n \geq 0}{\underrightarrow{\lim}} (Y_n, \phi_n)$, where ϕ_n denotes the inclusion isomorphism of $\pi_1(Y_n)$ in $\pi_1(Y_{n+1})$ ([33], p. 32), and

$\pi_1(Y_{-\infty}) = \underset{n \geq 0}{\underrightarrow{\lim}} (Y_{-n}, \rho_{-n})$, where ρ_{-n} denotes the inclusion isomorphism of $\pi_1(Y_{-n+1})$ in $\pi_1(Y_{-n})$ ([33], p. 32).

It follows then that $\pi_1(Y_0) \subset \pi_1(Y_\infty)$, and $\pi_1(Y_{-1}) \subset \pi_1(Y_{-\infty})$. This fact and the diagram above imply that

$$\pi_1(X) = \pi_1(Y_{-\infty} \cup Y_\infty) = \pi_1(Y_{-\infty}) \underset{\pi_1(S)}{*} \pi_1(Y_\infty) .$$

Note that if k is of genus g, then $\pi_1(S)$ is free of rank $2g$.

Suppose one of the maps, ${}_0f_i\colon \pi_1({}_0S_i) \to \pi_1(X_0)$ is not onto, say for $i=1$. Then no ${}_jf_i$ will be onto, so that

$$\pi_1(Y_\infty) \subsetneq \pi_1(Y_{-1} \cup Y_\infty) \subsetneq \pi_1(Y_{-2} \cup Y_\infty) \cdots$$

and

$$\pi_1(X) = \bigcup_{n=1}^{\infty} \pi_1(Y_{-n} \cup Y_\infty)$$

so that $\pi_1(X)$ is not finitely generated. But if ${}_0f_i\colon \pi_1({}_0S_i) \to \pi_1(X_0)$ is onto for $i = 1, 2,$ then all ${}_jf_i$ are onto so that

$$\pi_1(S) \approx \pi_1(X_0) \approx \pi_0(Y_1) \approx \pi_1(Y_n) \approx \pi_1(Y_\infty) \approx \pi_1(Y_{-1} \cup Y_\infty)$$

$$\approx \pi_1(Y_{-n} \cup Y_\infty) \approx \pi_1(Y_{-\infty} \cup Y_\infty) \approx \pi_1(X)$$

and $\pi_1(X)$ is free of rank $2g$. Hence if $[G, G] = \pi_1(X)$ is finitely generated, ${}_0f_1$ and ${}_0f_2$ are onto and $\pi_1(X)$ is free of rank $2g$. This proves the first assertion of Theorem 4.4.1.

If neither of the mappings ${}_0f_i$ is onto, then $\pi_1(X) = \pi_1(Y_{-\infty}) \underset{\pi_1(S)}{*} \pi_1(Y_\infty)$ is a proper free product with amalgamation and may be written as $\underset{n>0}{\underrightarrow{\lim}}\, \pi_1(Y_n \cup Y_{-n})$, where each $\pi_1(Y_n \cup Y_{-n})$ is a proper free product with amalgamation of $\pi_1(Y_n)$ and $\pi_1(Y_{-n})$ on a free group of rank $2g$ isomorphic to $\pi_1(S)$, so by virtue of equations 1 and 2, A) is proven. Suppose one of the maps, ${}_0f_i$, is onto, say ${}_0f_2$, and the other, ${}_0f_1$, is not, then

$$\pi_1(Y_\infty \cup Y_{-n}) \underset{\neq}{\subset} \pi_1(Y_\infty \cup Y_{-n-1})$$

and

$$\pi_1(S) \approx \pi_1(Y_0) \approx \pi_1(Y_1) \approx \pi_1(Y_n) \approx \pi_1(Y_\infty) \quad .$$

Since Y_∞ is homeomorphic to $Y_{-n} \cup Y_\infty$,[†] $\pi_1(Y_\infty) \approx \pi_1(Y_{-n} \cup Y_\infty)$, and hence each of the groups in the direct system,

$$\pi_1(Y_\infty) \underset{\neq}{\subset} \pi_1(Y_\infty \cup Y_{-1}) \underset{\neq}{\subset} \pi_1(Y_\infty \cup Y_{-2}) \cdots$$

is free of rank $2g$. Of course $\pi_1(X) = \underset{n>0}{\underrightarrow{\lim}}\, \pi_1(Y_\infty \cup Y_{-n})$. Since the mappings in this direct system are all inclusions, it follows that any finitely generated subgroup $H \subset \pi_1(X)$ lies in some $\pi_1(Y_\infty \cup Y_{-n})$ and so H is a subgroup of a free group and hence free [47]; in other words, $\pi_1(X) = [G, G]$ is locally free, and B) is proven. This completes the proof of the theorem.[‡] This theorem has many implications which will be found scattered throughout the remainder of the book. We state one corollary here which will be used in this chapter.

COROLLARY 4.5.1. The center of $[G, G]$ is trivial.

PROOF. If $[G, G]$ is finitely generated, it is free of rank > 1 according to Theorem 4.5.1, hence the corollary follows.

If $[G, G]$ is not finitely generated, it is either a free product with amalgamation on a centerless group and hence centerless according to [33] p. 32, or else locally free and non-abelian and hence centerless.

Q.E.D.

---

† The homeomorphism may be taken to be the covering translation t^n restricted to $Y_{-n} \cup Y_\infty$.

‡ The proof given above is identical to that given in [44].

Whether or not G' is finitely generated we may still look upon G' as in the picture below

$$\cdots \underset{F^{-2}}{*} A_{-1} \underset{F^{-1}}{*} A_0 \underset{F^0}{*} A_1 \underset{F^1}{*} A_2 \underset{F^2}{*} A_3 \underset{F^3}{} \cdots$$

A generator, t, of the group G/G' of covering translations acts upon G' by mapping A_i upon A_{i+1} by the natural isomorphism induced by the action of t on X_i.

Thus G may be described by giving the structure of G' as indicated in Theorem 4.5.1, since the automorphism of G' induced by conjugation by t coincides with the action of t on G' induced by a covering translation.

It is appropriate to mention here two results of Crowell [9] and Rapaport [53], and a result of Murasugi [40]. First, a theorem which will be proved in §9 of this chapter:

> THEOREM 4.5.2. (Rapaport-Crowell) If G' is finitely generated, then $|\Delta_G(0)| = 1$.

Here $\Delta_G(t)$ is the Alexander polynomial, an invariant to be described in §7.

Secondly, a partial converse to Theorem 4.4.1,

> THEOREM 4.5.3. (Rapaport-Crowell) If G' is free it is finitely generated.

Finally a theorem giving sufficient conditions for G' to be finitely generated.

> THEOREM 4.5.4. (Murasugi) If k is alternating and $|\Delta(0)| = 1$, then G' is finitely generated.

§6. The Alexander Matrix (Tentative Description)

In a paper [2] published in 1933, J. W. Alexander constructed over a certain ring, a matrix whose determinant, he proved, was up to a unit an invariant of the knot type of k. Since that paper, the matrix he constructed has been constructed by several other people by several different methods [20], [57], [41].

We shall give an interpretation of the Alexander matrix in this section, but the proof will be postponed.

Recalling Chapter II, $G/G' = |t: |$ acts on G'/G'', since $1 \to G'/G'' \to G/G'' \to G/G' \to 0$ is exact and G/G' is free, thus G'/G'' is a module over G/G'. It is clear that the structure of G/G'' is tied up with the module structure of G'/G''. We now assume:

A) The Alexander Matrix of k is a presentation [9] of $G/G'' \oplus G'/G''$ considered as a module over the integral group ring of $G/G' = |t: |$, where G/G' acts by multiplication on G/G' and by conjugation on G'/G''.

It will also be convenient to assume the following easily proved fact.

B) Suppose G, H are knot groups, then G'/G'' and H'/H'' are isomorphic as modules over the integral group ring of G/G' and H/H' respectively if and only if G/G'' and H/H'' are isomorphic.

We remark finally that the Alexander Matrix of a knot group G, may be computed from a finite presentation of G [20].

§7. The Alexander Polynomials

If A is an n columned Alexander Matrix for a knot group G, then the ideal, a, generated by the minors of rank $n-d$ is an invariant of G.[†] A generator of the smallest principal ideal containing a is

---

† It is proved in Zassenhaus, Theory of Groups, that given a matrix, these ideals are invariants of a module which the matrix may be assumed to present, and we have assumed that the Alexander Matrix, in fact, presents a module which is an invariant of the knot group. Accepting these facts makes most of this section academic.

called the dth Alexander Polynomial of G; it is determined only up to a unit, of course. The invariance of these polynomials is not obvious, nor is it trivial to prove. One proof depends upon the Tietze Theorem on equivalence of presentations of isomorphic groups [20]. To understand this proof one should appreciate the manner in which the Alexander Matrix may be constructed from a finite presentation of a group. We do not wish to develop the free calculus here, so that we must refer the reader to [20] and [19] for a full account of the construction of the Alexander Matrix. The proof goes roughly as follows:

1. Denote by G, H two knot groups, by JZ the group ring of the infinite cyclic group generated by t.

2. Suppose G and H are isomorphic.

3. There must exist a finite sequence of Tietze transformations [61] from a finite presentation of G to a finite presentation of H.

4. Tietze transformation I adjoins a new relation which is a consequence of the old relations. The effect on the Alexander Matrix of such an operation is to adjoin a new row which is a linear combination of old rows with coefficients units in JZ. This operation and its inverse leave invariant the ideal generated by the minors of rank n-d.

5. Tietze transformation II adjoins a new generator and a new relation setting this generator equal to some word in the old generators. The effect on the Alexander Matrix is to adjoin a new row, and a new column of zeros, except for the entry common to the new row and column, which will be a 1. Obviously, the ideal generated by the minors of rank n-d of such a matrix is unaffected by such an operation, or its inverse.

6. Numbers 3, 4, and 5, imply that the Alexander Polynomials of G are an invariant of G.

We shall subsequently indicate heuristically, how the Alexander Matrix is a presentation [9] of G'/G'' as a module over G/G'. If as agreed earlier we assume this last statement, an alternative proof of the invariance of the Alexander Polynomials may be given along the following lines. We use the same notation as the previously outlined proof.

1. $G'/G'' = M$ and $H'/H'' = N$ must be isomorphic as modules over the group ring of $G/G' \approx H/H'$ which we denote JZ.

2. If A and B are presentation matrices (Alexander Matrices) of M and N respectively, then the matrix, C, below presents M or N.

$$C = \begin{matrix} A & 0 \\ P & I \\ 0 & B \\ I & Q \end{matrix}$$

A, B have at least as many rows as columns. (Rows of zeros must be added if necessary.)

I is the identity matrix of the appropriate size.

0 is a matrix of zeroes of the appropriate size.

P describes the generators of the presentation of N in terms of the generators of M.

Q describes the generators of a presentation of M in terms of the generators of N.

3. Each row in the matrix $\begin{matrix} 0 & B \\ I & Q \end{matrix}$, is a linear combination of the rows of $\begin{matrix} A & 0 \\ P & I \end{matrix}$, and conversely.

4. Number 3 implies that the ideal generated by the minors of C of rank $n - d$ is equal to the ideal generated by the minors of rank $n - d$ of $\begin{matrix} 0 & B \\ I & Q \end{matrix}$ or $\begin{matrix} A & 0 \\ P & I \end{matrix}$.

So that ideal generated by minors of rank $r + s - d$ of $\begin{matrix} r & s \\ 0 & B \\ I & Q \end{matrix}$ = ideal generated by minors of rank $u + v - d$ of $\begin{matrix} u & v \\ A & 0 \\ P & I \end{matrix}$

and ideal generated by minors of rank $u - d$ of A = ideal generated by minors of rank $s - d$ of B,

and this implies the desired result.†

If A denotes a square matrix presenting G'/G'', then we contend:

PROPOSITION 4.7.1. Det A annihilates every element of G'/G''.

PROOF. With the same notation as before, we may consider the matrix A as defining an endomorphism of a free module F, and M as $F/ \operatorname{im} A$ [9]. Denote by φ the natural map from F to M.

If $x \in F$, then xA is in the kernel of φ. Recalling that $A(\operatorname{adj} A) = (\operatorname{adj} A)A = (\det A)I$, we see that $x(\det A)I = xA(\operatorname{adj} A) = x(\operatorname{adj} A)A = yA$, so that $x(\det A)I$ is in the kernel of φ, which implies that $\det A$ annihilates every element of M.

In a moment a much more general theorem of Crowell will be proved.

We shall postpone until §9 some theorems about the first Alexander Polynomial. In the next section we return again to consideration of the Alexander Matrix. At this point we present some of the high points of Crowell's investigations [9], [10], into the module structure of G'/G''. The proofs we give are Crowell's.

We begin by fixing notation. Jt denotes the integral group ring of G'/G''. A denotes the Jt module G'/G''. φ denotes the largest ideal in Jt which annihilates A. Δ_i denotes the ith Alexander Polynomial.

---

† A proof of this theorem may be found in [12].

When we say that the Alexander Matrix M, with entries a_{ij}, presents A, we mean that A is the factor module of the free Jt module generated by the columns, by the submodule spanned by the rows. More precisely we assume without loss of generality that M is square, and denote by X_1 the free Jt module generated by $x_1, \ldots, x_n$, and by X_0 the free Jt module generated by $r_1, \ldots, r_n$, then the mapping $d\colon X_0 \to X_1$ defined by

$$dr_i = \sum_{j=1}^{n} a_{ij}x_j$$

is a monomorphism and $e\colon X_1 \to X_1/dX_0 \approx A$.

We denote by φ_ℓ the ideal generated by the minors of M of order $n - \ell$. As we remarked earlier, these ideals are an invariant of G, and hence of A.

The following lemma will be used to prove G'/G'' is torsion free.

LEMMA 4.7.1. (Crowell) If $\pi a = 0$ for some prime $\pi \in Jt$ and $a \in A$ $a \neq 0$, then $\varphi_1 \subset (\pi)$.

PROOF.

$$a = e\left(\sum_{j=1}^{n} \alpha_j x_j\right)$$

for some α_j. Since $\pi a = 0$,

$$\pi\left(\sum_{j=1}^{n} \alpha_j x_j\right) = d\left(\sum_{i=1}^{n} \beta_i r_i\right) = \sum_{j=1}^{n} \pi\alpha_j x_j \quad ,$$

and

$$d\left(\sum_{i=1}^{n} \beta_i r_i\right) = \sum_{j=1}^{n}\left(\sum_{i=1}^{n} \beta_i a_{ij}\right) x_j \quad ,$$

by the definition of d. Thus

$$(\pi\alpha_j)X_j = \left(\sum_{i=1}^{n} \beta_i\alpha_{ij}\right) X_j$$

so that

$$\pi\alpha_j = \sum_{i=1}^{n} \beta_i\alpha_{ij} \quad . \tag{1}$$

Now consider $\varphi\colon Jt \to Jt/(\pi)$. For each j, $\varphi(\pi\alpha_j) = 0$, since $\varphi(\pi) = 0$, and so by (1) we have

$$0 = \sum_{i=1}^{n} \varphi(\beta_i)\varphi(\alpha_{ij}) \qquad \text{for each } j = 1, \ldots, n. \tag{2}$$

Looking upon (2) as a matrix equation

$$\left(\varphi(\beta_1) \cdots \varphi(\beta_n)\right)\left(\varphi(M)\right) = \begin{pmatrix} 0 \\ \vdots \\ 0 \end{pmatrix}$$

we conclude that either $\varphi(\beta_i) = 0$ $i = 1, \ldots, n$ or $\det(\varphi(M)) = 0$. The first alternative implies that

$$\pi\alpha_j = \pi \sum_{i=1}^{n} \gamma_j\alpha_{ij}$$

for $j = 1, \ldots, n$, but in this case

$$\sum_{j=1}^{n} \alpha_j X_j$$

lies in dX_0 so that $a = 0$ contrary to assumption. We are forced to the conclusion that $\det(\varphi(M)) = 0$; then $\varphi(\det M) = 0$ so that $\det(M) = \varphi_1 \subset (\pi)$, and the lemma is proved.

We apply this lemma directly to prove

THEOREM 4.7.1. (Crowell) A is torsion free.

PROOF. Suppose $pa = 0$ for p a prime in $Z \subset Zt$ and $a \neq 0$, $a \in A$. Then by Lemma 4.7.1 and the easily verifiable fact that p is a prime in Zt we conclude $(p) \supset \varphi_1 = (\Delta_1)$. But as we shall soon see $\Delta_1(1) = 1$ so that (p) cannot contain (Δ_1).

We complete our discussion of the current state of this particular aspect of Crowell's work by giving an interpretation of his proof of the following illuminating theorem.

THEOREM 4.7.2. (Crowell) $\varphi = (\Delta_1/\Delta_2)$.

PROOF. If Jt happened to be a principal ideal domain, then the theorem would follow from the standard diagonalization procedure which yields the structure theorem for finitely generated modules over a principal domain. Unfortunately Jt does not happen to be a principal ideal domain, but we may consider the entries of M to lie in Qt, which is a principal ideal domain. In order to carry over the desired result from Qt to Zt we need to know:

1. If $(\Delta_2)_{Zt}$ is the smallest principal ideal in Zt which contains the ideal φ_2, then $(\Delta_2)_{Qt}$ is the smallest (principal) ideal in Qt which contains φ_2;
2. The annihilator of A as a Qt module is the "same" as the annihilator of A as a Zt module.

This last condition should be taken with a few grains of salt, for as we shall see, the formalism used in this proof is somewhat different from the language of number 2 above.

Let us first deal with number 1. We recall that a polynomial in At is primitive if its coefficients are relatively prime. Denote by G the smallest ideal in Qt which contains φ_2. We wish to prove $G = (\Delta_2)_{Qt}$.

Obviously, we may assume $(\Delta_2)_{Zt} \neq 0$. Clearly it suffices to prove that any ideal in Qt containing φ_2, also contains $(\Delta_2)_{Qt}$. To this end let (c) denote an ideal in Qt which contains φ_2, and select a primitive generator $c \in Zt$. If $a \in \varphi_2$, then $a = \gamma c$ for some $\gamma \in Qt$. If $\gamma \notin Zt$ then since $a \in Zt$ some integer $s > 1$ divides the product of c and a primitive polynomial, but by Gauss' Lemma [4] the product of two primitive polynomials is primitive so that γ must lie in Zt. This means φ_2 is contained in the Zt ideal $(c)_{Zt}$. In Zt, $(\Delta_2)_{Zt}$ is the smallest principal ideal containing φ_2, so that $(\Delta_2)_{Zt} \subset (c)_{Zt}$, but this implies $(\Delta_2)_{Qt} \subset (c)_{Qt} = (c)$, so that $(\Delta_2)_{Qt}$ is contained in any ideal containing φ_2, and the proof of number 1 is complete.

In order to consider the entries a_{ij} as elements of Qt rather than Zt, it is convenient to tensor Qt with X_0 and X_1 and so obtain a matrix which presents $Qt \otimes_{Zt} A$ as a module over Qt. This last module has as annihilator, by virtue of the structure theorem for finitely generated modules over a principal ideal domain, and the just proven 1), the ideal generated by Δ_1/Δ_2.

We now need to make some sort of correspondence between the annihilator of A as a Zt module and A as a Qt module. This is most conveniently done by means of the following maps:

$$A \cong Zt \otimes_{Zt} A \quad \text{by} \quad a \to 1 \otimes_{Zt} a$$

so that

$$Q \otimes_Z A \cong Q \otimes_Z (Zt \otimes_{Zt} A)$$

and

$$Q \otimes_Z (Zt \otimes_{Zt} A) \cong (Q \otimes_Z Zt) \otimes_{Zt} A \cong Qt \otimes_{Zt} A$$

so that the map

$$g\colon\ Q \otimes_Z A \to Qt \otimes_{Zt} A \quad \text{defined by} \quad g(q \otimes_Z a) = q \otimes_{Zt} a$$

is an isomorphism. Defining

$$f\colon\ A \to Q \otimes_Z A \quad \text{by} \quad f(a) = 1 \otimes_Z a$$

and

$$h:\ A \to Qt \otimes_{Zt} A \quad \text{by} \quad h(a) = 1 \otimes_{Zt} a$$

we obtain the following consistent diagram:

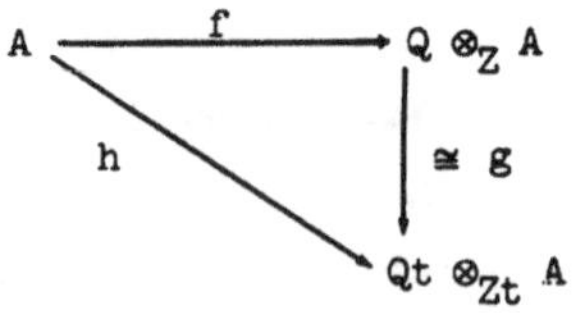

Since A is torsion free by Theorem 4.7.1, f and h are monomorphisms ([6], p. 130).

Now let us prove $(\Delta_1/\Delta_2) = \varphi$. If $a \in A$, then $\Delta_1/\Delta_2 \cdot h(a) = 0$ since Δ_1/Δ_2 annihilates $Qt \otimes_{Zt} A$. On the other hand $\Delta_1/\Delta_2 \cdot h(a) = \Delta_1/\Delta_2(1 \otimes_{Zt} a) = 1 \otimes_{Zt} \Delta_1/\Delta_2 a = h(\Delta_1/\Delta_2 a)$ so that $\Delta_1/\Delta_2 a = 0$ since h is a monomorphism. Thus $\Delta_1/\Delta_2 \in \varphi$. If $\alpha \in \varphi$ then $\alpha(\beta \otimes_{Zt} a) = (\beta \otimes_{Zt} \alpha a) = 0$ for $\beta \in Qt$, $a \in A$, so $\alpha \in (\Delta_1/\Delta_2)_{Qt}$ since we have proved $(\Delta_1/\Delta_2)_{Qt}$ is the annihilator of the Qt module $Qt \otimes_{Zt} A$. Thus we may write $\alpha = \gamma\Delta_1/\Delta_2$ for some $\gamma \in Qt$. Since Δ_1 is primitive, Δ_1/Δ_2 is primitive and we may apply Gauss' Lemma again to assert γ is in fact in Zt. Thus α is in $(\Delta_1/\Delta_2)_{Zt}$ and the proof is complete.

§8. The Alexander Matrix

We shall sketch in this section another description of the Alexander Matrix which we have found useful, and understandable.

We shall begin in a somewhat more general way than necessary by setting for ourselves the following problem.

We are given a finite number of generators x_i, and relations r_j for a group G. We suppose in fact that G is the fundamental group of a 3-manifold M, with a 1-dimensional subcomplex L removed, and our presentation has been obtained by means of a splitting complex for (M, L). We are given a homomorphism, φ, from G onto an abelian group H, with kernel K. The problem is to find a presentation for the group K/K', considered as a module over the group ring, JH, of $G/K = H$. We attack the problem as follows.

1. Construct the covering $\tilde{K}$ corresponding to K, and dualize its triangulation. Denote by S the splitting complex used, and by X, M split along S.

2. Note that $K/K' \approx H_1(\tilde{K}, Z)$; while K/K' looked upon as a group with operators in H, and $H_1(\tilde{K}, Z)$ operated on by the group H of covering translations, are isomorphic as groups with operators since $\tilde{K}$ is a regular covering, and H is abelian. This means K/K' and $H_1(\tilde{K}, Z)$ are isomorphic JH modules. We shall write hx to denote the operator $h \in H$ acting on a chain, cycle, or boundary in $\tilde{K}$, or $h \in H$ acting on an element $x \in H_1(\tilde{K}, Z)$.

3. A basis for the 2-chains of $\tilde{K}$ may be taken to be the 2-cells in $\tilde{K}$ lying over those 2-cells in M which are dual to the 1-simplices in $S-L$.

4. A JH basis for some of the 1-chains of $\tilde{K}$ may be taken to be the 1-chains lying over 1-chains in M which correspond to the generators x_i. We take as basepoint in $\tilde{K}$, a point in the sheet corresponding to the identity in H, and lying over the basepoint of M. A set of 1-chains $\hat{x}_i$ in 1-1 correspondence with the generators of G may then be selected.

5. A JH basis for the module of bounding 1-cycles in $\tilde{K}$ may be obtained in the following manner. Each r_j corresponds to a 2-cell c_j in $M-L$, we select the 2-cell $\tilde{c}_j$ in $\tilde{K}$ lying over c_j, with basepoint in the sheet of $\tilde{K}$ corresponding to the identity in H. The boundary of $\tilde{c}_j$ is read by lifting the boundary of c_j to $\tilde{K}$. This is done as follows: suppose $r_j = x_{i_1}$, then $\partial\tilde{c}_j = \hat{x}_{i_1}$. If $r_j = x_{i_1}^{-1}$ then $\partial\tilde{c}_j = [\varphi(x_{i_1}^{-1})]\,\hat{x}_{i_1}$. If $r_j = x_{i_1}x_{i_2}$, $\partial\tilde{c}_j = \hat{x}_{i_1} + [\varphi(x_{i_1})]\,\hat{x}_{i_2}$. If $r_j = x_{i_1}x_{i_2}^{-1}$, $\partial\tilde{c}_j = x_{i_1} - \varphi[x_{i_1}x_{i_2}^{-1}]\,\hat{x}_{i_2}$. In general if $r_j = wx_i$, $\partial\tilde{c}_j = A+[\varphi(w)]\,\hat{x}_i$ if

$r_j = wx_i^{-1}$, $\partial\tilde{c}_j = A - [\varphi(wx_i^{-1})]\,\hat{x}_i$ where A is what $\partial\tilde{c}_j$ would be if $r_j = w$.

The following picture may make this lifting process clear.

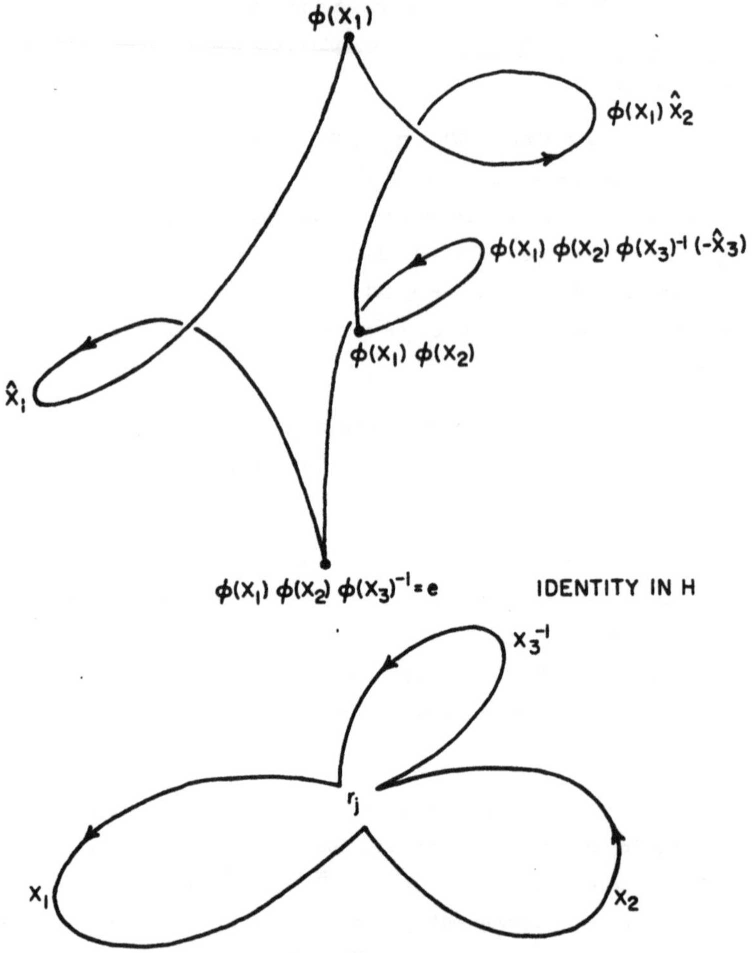

Fig. IV-1

Thus

$$\partial r_j = \sum_{i=1}^{n} a_{ij}\,\hat{x}_i\ , \qquad a_{ij} \in JH.$$

6. Construct a matrix, B, each row corresponding to a relation, each column corresponding to a chain $\hat{x}_i$, and the entry in the ith row and j^{th} column being a_{ij}. The row space of this matrix is the submodule of boundaries in the module of 1-chains. (This matrix is the Jacobian of G evaluated at φ [11].)

7. Consider now the specific situation of a knot group G. (With a small amount of fiddling) the presentation we obtained in Chapter III may be brought into the following form:

$$(t, x_1, x_2, \ldots, x_n: \ t\,w_j t^{-1} = v_j) \qquad j = 1, 2, \ldots, n$$

where w_j, v_j are words in the x_i, and the x_i lie in G'. It follows then that the smallest normal subgroup containing the x_i is G'.

8. We let K = G', so that H ≈ Z. It is easy to see that the $\hat{x}_i$ are generators for the JH-module of 1-cycles in $\tilde{K}$. Furthermore, since the 1-chains of $\tilde{K}$ generate a free JH-module, the JH-module of 1-cycles in $\tilde{K}$ is free, and the $\hat{x}_i$ are a free basis.

9. It follows from numbers 6 and 2 that the matrix B, with the column corresponding to t removed, is a presentation of the JH-module G'/G".

10. The Alexander Matrix is B.

At this point we must recognize a mild contradiction which has occurred. B <u>is</u> an Alexander Matrix for G but it is really B with the column corresponding to t <u>removed</u> which represents the JZ-module G'/G". We also note that the column corresponding to t has nothing but zeros in it, so that the "Alexander Polynomial" is the same whether or not this column is included. Furthermore, with this column included, B presents the JH-module JZ ⊕ G'/G" as previously assumed.

This interpretation of the Alexander Matrix is not meant as a definition. The definitions made in the literature which suggest this way of looking at the Alexander Matrix are to be found in Fox [19], Lyndon [35] and Trotter [63].

§9. The Alexander Polynomial

In this section we shall indicate some uses to which the Alexander Polynomial may be put in the investigation of a knot group.

We will make a slight change of notation now; if G is a knot group, we denote by $\Delta_G(t)$ the first Alexander Polynomial of G.

THEOREM 4.9.1. $\Delta_G(t) = 1$ if and only if G'/G'' is trivial.

PROOF. By Proposition 4.7.1, the interpretation we have given to the Alexander Matrix, and its squareness, $\Delta_G(t)$ annihilates every element of G'/G''. If 1 annihilates every element of G'/G'' then G'/G'' is trivial. If G'/G'' is trivial it is presented as a module over JZ by a matrix with determinant equal to 1. This proves the theorem.

THEOREM 4.9.2. If G has a center then every root of $\Delta_G(t)$ is a root of unity.

PROOF. We require a theorem whose proof we postpone untill Chapter V, (Theorem 5.4.3). The theorem states that if G has a center, $Z(G)$, then G' is free of finite rank. Assuming this, the proof proceeds as follows.

For knot groups whose commutator subgroups are free, the Alexander Matrix (as in numbers 9 and 10 in §8 of this chapter) is simply $M - tI$ where M is a unimodular integer matrix which describes the automorphism of G'/G'' induced by a generator t of G/G' and I is the identity matrix. Since $Z(G) \cap G' = 1$ by Corollary 4.5.1, it follows that if $Z(G) \neq 1$, some power say r, of t leaves each element G'/G'' fixed, so that M^r is I, hence every characteristic root of M is an

r^{th} root of unity. But $|M - tI|$ is just the Alexander Polynomial, and so the theorem is proved.

THEOREM 4.5.2. (Rapaport) If $\Delta_G(0) \neq 1$ then G' is not finitely generated.

The following proof is not that given in [53], but we present it in this manner since it will prove useful for some later considerations.

PROOF. If G' is finitely generated, then G' is by Theorem 4.5.1 free, so that G'/G" is a free abelian group. Since t induces an automorphism of this free abelian group, the action of t on G'/G" may be described in the standard way by a square, integral, unimodular matrix, M. It follows that an Alexander Matrix giving a presentation of G'/G" as a module over JZ may be taken to be M - tI. But det (M-tI) = $\Delta_G(t)$ so that $\Delta_G(0) = 1$ since M is invertible.

THEOREM 4.9.4. (Seifert) $\Delta_G(t) = t^{\lambda} \Delta_G(t^{-1})$.

PROOF. This is a deep property of the Alexander Polynomial of a knot group, and has been proven in several quite different ways. (Seifert [57], Blanchfield [5], Fox [62] and Milnor [37].) The reader is referred to any of these for proof. The last two proofs mentioned are particularly interesting.

THEOREM 4.9.5. (Seifert) $\Delta_G(1) = 1$.

PROOF. The Alexander Matrix presents the JZ-module JZ ⊕ G'/G". We contend that if t is mapped onto 1 then G'/G" becomes a trivial J module. This is true because G'/G" is destroyed by mapping the generator t to 1 in the presentation for G, $(t, x_1, x_2, \ldots, x_n: t w_j t^{-1} = v_j)$ $j = 1, \ldots, n$, and G/G" itself, is described by presenting G'/G" as a module over G/G" ≈ Z. If G'/G" did not become a trivial J-module when t was mapped to 1, G/G" would not become a trivial group when t was mapped to 1.

Since G'/G'' is a trivial J-module when t is mapped to 1, it follows that its presentation matrix must represent the trivial module if t is mapped to 1, so that indeed $\Delta_G(1) = 1$.

THEOREM 4.9.6. (Seifert) If

$$\rho(t) = \sum_{i=0}^{n} a_i t^i$$

satisfies a) $\rho(1) = 1$ and
b) $\rho(t) = t^n \rho(t^{-1})$
then there exists a knot group G such that $\Delta_G(t) = \rho(t)$.

PROOF. The proof may be found in Seifert's paper [57]. It involves a study of the Alexander Matrix as it is constructed from an orientable surface spanning a knot.

We shall illustrate other useful properties of the Alexander Polynomial in other chapters.

CHAPTER V

SUBGROUPS

§1. Introduction

In this chapter a number of results will be proved or cited, which indicate some properties of subgroups of knot groups. In view of the preceding chapter, the subgroups to consider first would seem to be those which arise as the kernel of the natural map of a knot group onto a cyclic group of finite order. These subgroups have importance in another context as well. According to P. A. Smith, the fixed point set of a periodic map of S^3 on itself is a sphere, (possibly -1 dimensional). It is natural to ask whether such a mapping can admit a knotted curve as its fixed point set. This problem is as yet unresolved (in dimension 3). If such a mapping exists, then the orbit space is a simply connected 3-manifold containing a knot, the 3-sphere minus the original knot thus is a finite cyclic covering of the orbit space minus the new knot. This leads to the algebraic question: Can one knot group be isomorphic to the subgroup K_n of another knot group?[†]

§2. Kernels of Maps to Z_n

The structure Theorem 4.5.1 of Chapter IV suggests a similar theorem for finite cyclic coverings. We shall prove such a theorem.

First let us see how the covering corresponding to the kernel K_n of the map from a knot group G onto Z_n may be constructed.

[†] Giffen [26] has answered this question in the affirmative.

If $\tilde{X}$ denotes the covering corresponding to G', then $\tilde{X}/\varphi^n$ = (the orbit space under the action of the nth power of a generator of the group of covering translations of $\tilde{X}$) = (the covering corresponding to K_n) = $\tilde{X}_n$.

Clearly, $\tilde{X}_n$ may be obtained from n copies of S^3 split along an orientable surface S of minimal genus spanning k, and matched together along their boundaries as in the figure below.

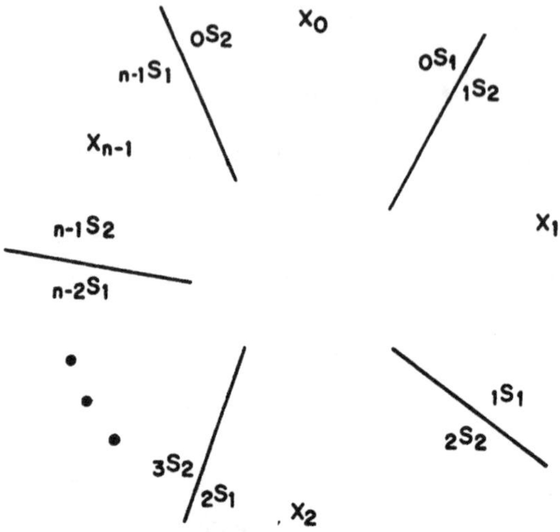

The Notation is that of Chapter IV

Fig. V-1

Application of Lemmas 4.4.1 and 4.4.2, and the van Kampen Theorem to the situation just described yields the following theorem.

THEOREM 5.2.1. If G' is finitely generated, then $K_n \approx (Z * G')/u(G')u^{-1} = \varphi^n(G')$ where φ is induced by the action of a generator of G/G' pulled back to G and acting on G' by conjugation (φ is induced by a generator of the group of covering translations, and so is an automorphism), u is a generator of Z, and G' is free.

If G' is a non-trivial free product with amalgamation, then

$$K_n \approx \frac{[Z * (\pi_1(X_1) \underset{\pi_1({}_1S_1)}{*} \pi_1(X_2) \underset{\pi_1({}_2S_1)}{*} \cdots \underset{\pi_1({}_{n-1}S_1)}{*} \pi_1(X_n)]}{u(\pi_1({}_0S_1))u^{-1}} = \varphi^n({}_0f_2(\pi_1({}_0S_1))$$

where $\pi_1({}_0S_1)$ is imbedded in $\pi_1(X_1)$ by the map ${}_1f_1$: $\pi_1({}_0S_1) \to \pi_1(X_1)$. (The notation is that of Theorem 4.5.1.)

A generator of Z is denoted u, and φ again is induced by a generator of the group of covering translations. Note that $\varphi^n({}_0f_2(\pi_1({}_0S_1)) \subset \pi_1(X_n)$, and φ is 1-1.

If G' is a direct limit of free groups, then $K_n \approx (Z * \pi_1(X_1))/u(\pi_1(X_1))u^{-1} = \varphi^n(\pi_1(X_1))$.[†‡]

While the situation as it is described in this theorem is not particularly simple, the first homology group of the covering corresponding to K_n, i.e., $K_n/[K_n, K_n]$ may be very easily described.

§3. $K_n/[K_n, K_n]$

Suppose M denotes a relation matrix for G'/G'' as a module over G/G', i.e., an Alexander Matrix with a column of zeros removed. To compute a presentation for G'/G'' one must consider the relations of G'/G'' qua module in terms of the generators of G'/G'' considered as a group. These generators are clearly $t^\alpha x_i$ where $\alpha = \ldots, -1, 0, 1, \ldots$ $i = 1, \ldots, m$ and the x_i generate G'/G'' considered as a module. Thus,

---

† The u in each of these three situations arises from a loop representing a generator of G/G' raised to the nth power. Such a loop represents, of course, an element of K_n.

‡ See the first footnote, Chapter IV, §5.

a typical relation from M, $(2t^3 + t^{-1})x_1 - (t^2 - 1)x_2$ considered as a relation in G'/G'' would lead to relations of the form:

$$2t^2x_1 + t^{-2}x_1 - tx_2 + t^{-1}x_2 = 0 \quad ,$$

$$2t^3x_1 + t^{-1}x_1 - t^2x_2 + x_2 = 0 \quad ,$$

$$2t^4x_1 + x_1 - t^3x_2 + tx_2 = 0, \quad \text{etc.},$$

and a relation matrix for G'/G'' would a priori have $m \cdot \infty$ columns, and $m \cdot \infty$ rows. On the other hand, the same relations as those above lead, by virtue of the collapsing under, say, φ^2, to the following relations in $K_2/[K_2, K_2]$ (where t^r and t^{r+2} are identified since $t^2x_i = x_i$ in the covering corresponding to K_2).

$$2x_1 + x_1 - tx_2 + tx_2 = 3x_1 = 0 \quad ,$$

$$2tx_1 + tx_1 - x_2 + x_2 = 3tx_1 = 0 \quad ,$$

$$2x_1 + x_1 - tx_2 + tx_2 = 3x_1 = 0 \quad .$$

It is clear now how to pass to the general case, so that a relation matrix for $K_n/[K_n, K_n]$ will have a priori $m \cdot n$ rows and $m \cdot n$ columns. This matrix will have one column for each $t^\alpha x_i$ $\alpha = 0, 1, \ldots, n-1$ $i = 1, \ldots, m$, and one row corresponding to each row of M, multiplied by t^α. In addition, there will be a column of zeros resulting from the presence of u as a generator of K_n (Theorem 5.2.1). A small amount of consideration of this situation yields,

THEOREM 5.3.1. (Fox [21], Zariski, Burau) A relation matrix for $K_n/[K_n, K_n]$ may be obtained from M by substitution of

$$\overbrace{\begin{matrix} 0 & 1 & 0 & \cdots & 0 \\ 0 & 0 & 1 & \cdots & 0 \\ & \cdots & & \cdots & \\ 0 & 0 & 0 & \cdots & 1 \\ 1 & 0 & 0 & \cdots & 0 \end{matrix}}^{n} = L$$

for each t in M, and S times the $n \times n$

identity matrix for each integer S in M, and then adjoining a column of zeros.

In the case where G' is finitely generated, Theorem 5.3.1 yields a simpler result.

THEOREM 5.3.2. A relation matrix for $K_n/[K_n, K_n]$ when G' is finitely generated is $M^n - I$.

PROOF. φ is described by M.

From this follows immediately,

COROLLARY 5.3.3. If G' is finitely generated and all roots of the Alexander Polynomial are ν^{th} roots of unity, then the groups $K_n/[K_n, K_n]$ and $K_{n+\nu}/[K_{n+\nu}, K_{n+\nu}]$ are isomorphic.

PROOF. $\Delta_G(t) = \det(M - tI)$, but by the Cayley-Hamilton Theorem $\Delta_G(M) = 0$ so that M is a root of $t^\nu - 1 = 0$ since $\Delta_G(t)$ divides $t^\nu - 1$.

Thus $M^{\nu+1} = M$. Since $M^n - I$ is a relation matrix for $K_n/[K_n, K_n]$ the corollary is proved.

Returning to the more general situation, use may be made of Theorem 5.3.1 to determine the betti number of $K_n/[K_n, K_n]$, and the order of the torsion part of this abelian group.

THEOREM 5.3.3 (Fox [21], Zariski, Burau) The betti number of $K_n/[K_n, K_n]$ is equal to 1 plus the number of nth roots of unity which are roots of $\Delta_G(t)$. The order of the torsion subgroup of $K_n/[K_n, K_n]$ is equal to

$$\prod_{i=1}^{n} \Delta_G(\zeta^i) / \prod_{j=1}^{\ell} (\zeta^{s_j})^{u_j}$$

where ζ is a primitive n^{th} root of 1,

each ζ^{s_j} is a root of multiplicity μ_j of $\Delta_G(t)$ and ℓ is the number of distinct n^{th} roots of $\Delta_G(f)$.

PROOF. By virtue of Theorem 5.3.1 the first part of the theorem should follow from a computation of the nullity of the matrix presenting $K_n/[K_n, K_n]$. This computation may be made as follows. Without altering the nullity of M, we may find matrices P, R over the rational group ring QZ such that $P\ M\ R$ is diagonal. We then obtain a matrix

$$\begin{pmatrix} \Delta_1(t) & \cdots & & 0 \\ & \Delta_2(t) & & \\ & & \ddots & \\ 0 & \cdots & & \Delta_m(t) \end{pmatrix}$$

where

$$\prod_{i=1}^{m} \Delta_i(t) = \Delta_G(t) \quad .$$

Now substitution of L for t in M, and sI for integers s in M yields the presentation matrix, N, of $K_n/[K_n, K_n]$. If $p_{ij}I$, and $r_{ij}I$ are substituted for the entries p_{ij}, r_{ij} in P and R respectively, then it is easy to see that the resulting matrices $P(I)$, $Q(I)$ have the property that

$$P(I)\ N\ Q(I) = \begin{pmatrix} \Delta_1(L) & & \cdots & 0 \\ & \Delta_2(L) & & \\ & & \ddots & \\ 0 & \cdots & & \Delta_m(L) \end{pmatrix}$$

This in turn implies that the nullity of N is equal to the sum of the mullities of the $\Delta_i(L)$. It is possible to diagonalize L over the complex numbers by means of the matrix $w = \zeta^{\alpha\beta}$ $\alpha, \beta = 1, \ldots, n$ and ζ a primitive n^{th} root of the unity, and $w^{-1} = \frac{1}{n}\zeta^{-\alpha\beta}$. The resulting matrix is

$$W^{-1}LW = \begin{pmatrix} \zeta & \cdots & & 0 \\ & \zeta^2 & & \\ & & \ddots & \\ 0 & \cdots & & \zeta^n \end{pmatrix}$$

It follows then that $W^{-1}\Delta_i(L)W = \Delta_i(W^{-1}LW)$ hence the nullity of $\Delta_i(L)$ is equal to the number of nth roots of unity which are roots of $\Delta_i(t)$, but since

$$\prod_{i=1}^{n} \Delta_i(t) = \Delta_G(t)$$

the roots of $\Delta_i(t)$ are roots of $\Delta_G(t)$, and the first part of the theorem is proved.

The order of the torsion subgroup of $K_n/[K_n, K_n]$ may be calculated from $W^{-1}\Delta_i(L)W$. It is clearly

$$\prod_{j=1}^{n} \prod_{i=1}^{n} \Delta_i(\zeta^{t_j})$$

where $t_j = 0$ if ζ^{t_j} is a root of $\Delta_G(t)$, and $t_j = j$ otherwise. The second part of the theorem now follows directly.[†]

The structure of $K_n/[K_n, K_n]$ may be computed for a particular knot by means of an orientable surface spanning the knot. For a fine account of this method, first devised by Seifert [57], see [18]. One result of the application of this technique is

THEOREM 5.3.4. (A. Plans[‡]) For any $n \geq 0$ $K_{2n+1}/[K_{2n+1}, K_{2n+1}] \approx Z \oplus Z_r \oplus Z_r$, for some r.

Other results for genus 1 knots will be found in [22].

[†] The proof given here is exactly that given by Fox in [21].

[‡] Aportación al estudio de los gruppos de homologiá de los recubrimientos cíclicos ramificados correspondientes a un nudo. Rev. Acad. Ci. Madrid 47 (1953), pp. 161-193.

Having investigated all the normal subgroups whose factor groups are abelian, let us turn now to study abelian subgroups.

§4. Abelian Subgroups

Of fundamental importance in the study of knot groups is the following theorem.

> THEOREM 5.4.1. (Papakyriakopoulos [51]) The complement of a knot is aspherical. This theorem may also be expressed by saying: the complement of a knot has the homotopy type of a $K(G, 1)$ space.

This latter statement along with the easily verified fact that the complement of a knot contains a 2-dimensional complex as a deformation retract implies that the homology groups of G with any coefficients vanish above dimension 2. We shall have occasion to apply this fact more generally later in this chapter. For the moment we may apply it to the study of the abelian subgroups of a knot group.

> THEOREM 5.4.2. (Papakyriakopoulous [51], Conner [7]) If $A \subset G$, and A is abelian and non-trivial, then A is either infinite cyclic, or the direct sum of two infinite cyclic groups.

PROOF. By virtue of the remarks above, the covering $\widetilde{A}$ corresponding to A is of the same homotopy type as a 2-dimensional $K(A, 1)$ space.

It follows then that A cannot be of finite order, as a finite abelian group has homology in an infinite number of dimensions. Similarly, if $A \supset Z \oplus Z \oplus Z$, then $H_n(A, Z) \approx Z$ for some $n \geq 3$, which is impossible since $\widetilde{A}$ is of the same homotopy type as a 2-dimensional complex.

Theorem 5.4.2 reduces the possibilities for abelian subgroups of a knot group to Z, and $Z \oplus Z$. Both these groups in fact occur, the first being generated, of course, by any non-trivial element, and the second being found on a thin torus containing the (non-trivial) knot in its interior.

This section is concluded with a discussion of the center of a knot group. Recalling Theorem 4.9.2, we begin with a theorem relating the existence of a non-trivial center to the structure of the knot group.

THEOREM 5.4.3. If G has a non-trivial center, then $[G, G]$ is free of finite rank.

PROOF. Recall that Theorem 4.5.1 proves that $[G, G]$ may take one of three forms. One form satisfies the conclusion of the theorem. The other two will be shown to be impossible.

According to Corollary 4.5.1, the center of G, $Z(G)$ does not intersect $[G, G]$, consequently the abelianizing homomorphism maps $Z(G)$ isomorphically onto a subgroup of $Z = |t: \ |$. Let t^r denote a generator of this subgroup. Since Z is a free group, G splits and every element may be written in the form $t^s C$, where $C \in [G, G]$. Let $t^r C$ denote a generator of $Z(G)$. Then $(t^r C) X (t^r C)^{-1} = X$ so that

$$(1) \qquad t^{-r} X t^r = C X C^{-1} \quad ,$$

for all $X \in G$.

Suppose now that $[G, G]$ has the form

$$\cdots F_{2g}^{*} \, A_{-1} \, \overset{*}{F}_{2g} \, A_0 \, \overset{*}{F}_{2g} \, A_1 \, \overset{*}{F}_{2g} \cdots$$

(recall, each factor of a free product with amalgamation is a subgroup of that product). Let X denote an element of A_0 which is not in $A_0 \cap A_1$.

Now the action of t on $[G, G]$ is to map each A_i onto A_{i+1}, so that $t^{-r} X t^r$ lies in A_{-r}, and since

$$X \notin A_0 \cap A_1, \quad (t^{-r} X t^r) \notin A_{-r} \cap A_{-r+1} \quad ;$$

it follows then that $(t^{-r} X t^r) \notin A_{-1} \cap A_0$ hence $t^{-r} X t^r \notin A_\infty$, but

$C X C^{-1} \in A_\infty$ so that (1) cannot be true. This dispenses with possibility A in Theorem 4.5.1. Case B is treated similarly as follows: if $[G, G] = A_0 \subset A_1 \subset A_2 \subset \ldots$ and $C \in A_k$, then pick $X \in A_k$, then $(t^r X t^{-r}) \in A_{k+r}$ $(t^r X t^{-r}) \notin A_{k+r-1}$ but $C X C^{-1} \in A_k$ and since $r > 0$, Equation (1) is impossible. This completes the proof of the theorem.

As far as the structure of the center is concerned, Corollary 4.5.1 implies that the abelianizing homomorphism mapping G onto Z is a monomorphism when restricted to the center of G, thus we may assert;

THEOREM 5.4.4. The center of a knot group is cyclic.

Finally for convenience we may restate Theorem 4.9.2: If G has a center, every root of $\Delta_G(t)$ is a root of unity.

§5. Homology of Subgroups

We simply wish to point out here a consequence of the fact that the complement of a knot has the homotopy type of a 2-dimensional K(G, 1) space, namely

THEOREM 5.5.1. The homology of a subgroup of a knot group vanishes in dimensions greater than 2.

PROOF. Any subgroup, A, of a knot group G, corresponds to a 2-dimensional covering of a 2-dimensional K(G, 1), this covering has the homotopy type of a K(A, 1), and since it is 2-dimensional, the theorem follows.

§6. Commutator and Central Series

Of great interest in the study of many classes of groups, the commutator series seems to be of limited value in the study of knot groups. The reason for this becomes apparent when one considers that the commutator subgroup may either be free (Theorem 4.5.1) or perfect (Proposition 4.7.1 and Theorem 4.9.6). This state of affairs does not however, preclude some interesting questions concerning this series (see Chapter XI).

The lower central series, $G^{(i)}$, [33] of a knot group is uninteresting since $G^{(1)} = G^{(2)}$. This follows from the fact that $G/G^{(1)} = G/[G, G] \approx Z$. One argument (to be found in [71]) runs as follows.

$$[G/G^{(2)}, G^{(1)}/G^{(2)}] = [G, G^{(1)}]/G^{(2)} = 0$$

hence

$$G^{(1)}/G^{(2)} \subset Z(G/G^{(2)}) .$$

Noting that $G^{(2)} \supset G^2$ we see that $G^{(1)}/G^{(2)}$ is abelian and so the above $G/G^{(2)}$ is a central extension of the abelian group $G^{(1)}/G^{(2)}$ by the cyclic group $G/G^{(1)}$. This implies $G/G^{(2)}$ is abelian so $G^{(2)} \supset G' = G^{(1)}$. Since $G' = G^{(1)} \supset G^{(2)}$ the assertion is proved.

§7. Ends

In a later chapter we shall discuss the matter of the number of ends [25] of a knot group. For the present chapter, the following theorem is pertinent. Suppose H is a subgroup of G.

THEOREM 5.7.1. If $[G: H] < \infty$ then H has one end.

PROOF. According to [52] G has one end, by [29] H has one end.

CHAPTER VI

REPRESENTATIONS

§1. Introduction

It is with some regret that we note the brevity of this chapter. While the representation of a given knot group on some non-trivial group seems to require mostly patience, there are almost no general theorems regarding the existence of finite representations. In this regard, we feel that the central problem is one first asked of the author by S. Abhyankar. We mention it below and again in the section on problems;

> QUESTION. Is every knot group residually finite? That is, given any $g \in G$ $g \neq 1$, does there exist a finite group F, and a homomorphism $\varphi\colon G \to F$, such that $g \notin$ kernel φ?

A fractional answer to this question is provided in this chapter, but the problem remains far from settled.

We begin what little we have to say with some results of Fox on metacyclic representations [18].

§2. Metacyclic Representations

Recall that a metacyclic group, M, is defined by a presentation; $(\tau, \varpi;\ \tau^q = 1 = \varpi^p\ \tau\varpi\tau^{-1} = \varpi^k)$, $k > 1$, $k^q \equiv 1(p)$, $(k-1, pq) = 1$. It is easy to see that M' is cyclic, generated by ϖ, and $M/M' \approx Z_q$.

Suppose we wish to represent the knot group G in M. Assume, as in Chapter IV, $G = (t, x_1, \ldots, x_n: r_1, \ldots, r_n)$, where t maps into a generator of G/G', and the $x_i \in G'$. By the last paragraph it is no loss of generality to attempt to map t to τ, and each x_i to some power of ω. Since the Alexander Matrix describes G/G'' (Chapter IV), and we are attempting to map G into a group M, with $M'' = 1$, our efforts to map G into M will utilize the Alexander Matrix of G.

Each row of the Alexander Matrix represents a linear combination of conjugates of the x_i by various powers of t. A map of G into M will be defined if we send t to τ and x_i into such powers s_i of ω, that the image of each linear combination of conjugates of the x_i by the various powers of t described by a row of the Alexander Matrix, maps into a trivial element of the cyclic group generated by ω. More precisely, if $\Sigma_i \; p_{ij}(t)x_i$ describes the jth row of the Alexander Matrix,† then

$$\prod_i \omega^{p_{ij}(k)s_i} = \omega^{\Sigma_i \, p_{ij}(k)s_i}$$

must equal 1 in M, for each j, if a homomorphism is to be defined by mapping x_i to ω^{s_i}, and t to τ. This implies $\Sigma_i \, p_{ij}(k)s_i \equiv 0 \; (p)$ for each j. It follows that the determinant of the Alexander Matrix evaluated at k must be congruent to 0 modulo p. Thus we have:

THEOREM 6.2.1. (Fox) A necessary condition that G be representable non-trivially in M is that $\Delta_G(k) \equiv 0 \; (p)$.

The proof of this theorem given above yields a technique for finding representations of G in M. One need only solve the n congruences

(1) $$\sum_{i=1}^{n} p_{ij}(k)s_i \equiv 0 \; (p) \quad , \quad j = 1, 2, \ldots, n .$$

† Recall that an entry t in the ith column of the Alexander Matrix corresponds to the element tx_it^{-1} in G/G''.

Then each solution set s_i $i = 1, 2, \ldots, n$ defines a homomorphism by mapping t to τ and x_i to ω^{s_i}.

§3. Non-trivial Representations of Non-trivial Elements

As mentioned in the introduction to this chapter, one of the central problems in the study of the representations of a knot group is that of the intersection of the normal subgroups of finite index. As might be expected, a knot group whose commutator subgroup is finitely generated proves to be a tractable object from this point of view. In fact, we have the following theorem.

THEOREM 6.3.1. If G' is finitely generated, then the intersection of all normal subgroups of finite index in G is the identity.

PROOF. It is clearly equivalent to prove that given any $g \in G$, $g \neq 1$, there exists a homomorphism φ, from G to a finite group F, such that $\varphi(g) \neq 1$.

If $g \notin G'$ then $\psi\colon G \to G/G' \approx Z$ maps g to a non-zero power, α of a generator, t, of G/G', and then $\rho\colon Z \to Z_{\alpha+1}$ composed with ψ maps G onto a finite cyclic group, and $\rho\psi(g) \neq 1$, so that we may assume $g \in G'$. Now according to Theorem 4.5.1, G' is free, and according to a well known result, [33], [30] or [45], any free group may be mapped onto a finite group so that g is not in the kernel, L_0, of this mapping. Let us denote by $L_1, L_2, \ldots, L_n$ the images of L_0 under the action of G/G' on G'. There are only a finite number of these images, since a finitely generated group has only a finite number of normal subgroups of a given finite index. Since the intersection of a finite number of subgroups of finite index is again of finite index, $\cap_{i=0}^{n} L_i = L$ is of finite index in G'. By the construction of L, G/G' acts on G' in such a way that L is mapped onto itself. This implies that G may be mapped onto G/L by a homomorphism θ, and we have the following situation

$$0 \to G'/L \to G/L \to G/G' \to 0 \quad ,$$

where G'/L is a finite group, and $\theta(g) \neq 1$, because $g \notin L_0$.

Since G'/L is finite, some power, say ν, of t, a generator of G/G', leaves G'/L fixed. This enables us to map G/L onto $\frac{G}{L} / t^\nu$ by a homomorphism η. Clearly, $\eta\, \theta(g) \neq 1$ since $t^{n\nu} \notin G/L$ for any $n > 0$, and $\theta(g) \neq 1$. This completes the proof of the theorem.

§4. The Range of Finite Homomorphs

It is of some interest, from both a topological and algebraic point of view, to know what finite groups are homomorphs of knot groups. The former stems from investigations of the groups of 2-spheres in the 4-sphere, and the latter from the desire to capture the algebraic properties of the class of knot groups.

Aside from some rather nugatory statements which follow trivially from previous results (e.g., $Z_2 \oplus Z_2$ is not a homomorph of a knot group since $G/G' \approx Z$, and Z cannot be mapped onto $Z_2 \oplus Z_2$) the only results seem to be the following. We denote by S_n and A_n, the full symmetric and alternating groups of degree n.

THEOREM 6.4.1. For every n, S_n is a homomorph of some knot group.

THEOREM 6.4.2. For every n, A_{2n+1} is a homomorph of some knot group.

Both these theorems follow from the particular knots and representations shown in Figs. VI-2 and VI-3. A permutation associated with an overpass corresponds to mapping the element of π_1 as shown in Fig. VI-1 onto that permutation in S_n. That these assignments define a homomorphism may be verified by examining the product of permutations arising from the relations at the crossing points.

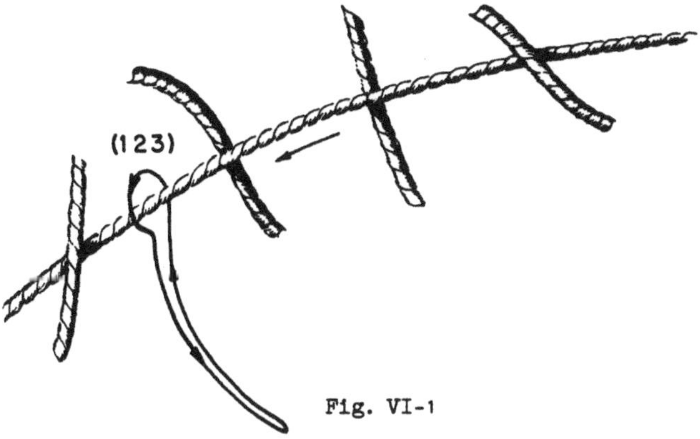

Fig. VI-1

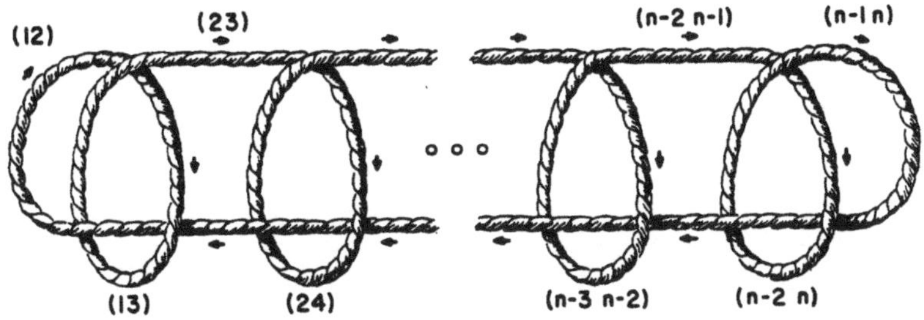

Fig. VI-2

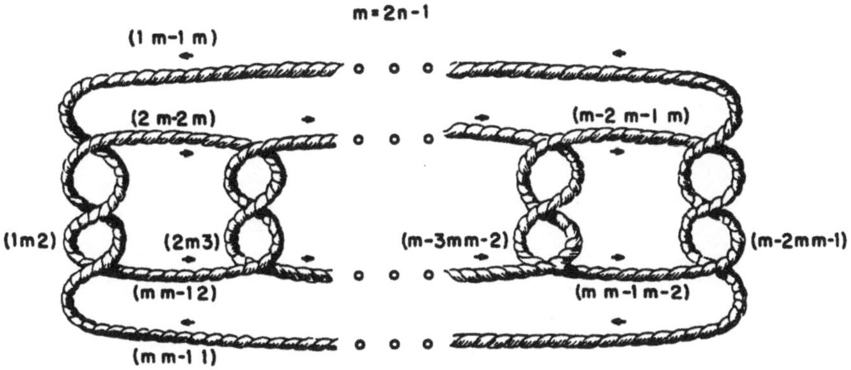

Fig. VI-3

CHAPTER VII

AUTOMORPHISMS

§1. Introduction

It is to be expected that the automorphisms of a knot group are intimately tied to the geometry of the knot, and in fact this is the case. However, the geometry is not entirely captured by the group alone. We shall define geometrically certain conjugate elements, and subgroups which hold useful information.

The boundary of a small neighbourhood of the knot is a torus, T, whose fundamental group is $Z \oplus Z$. If this torus is connected by an arc in S^3-k to a base point in S^3-k, then there is a natural mapping from $\pi_1(T)$ into $G = \pi_1(S^3\text{-}k)$. This mapping is a monomorphism if k is knotted [13], [51]. The subgroup of G so obtained depends on the arc joining T to the basepoint, but different arcs define subgroups which are conjugate in G. This conjugacy class (all conjugates are obtainable) has been referred to as the <u>group system</u> [15] of the knot group, and a member of this class as a peripheral subgroup. Selections of an orientation of S^3, and an orientation of k define two elements uniquely in each peripheral subgroup. This is accomplished as follows. $H_1(T)$ maps into $H_1(S^3\text{-}k)$ by ι, which is induced by the natural inclusion, and $\pi_1(T)$ maps into $H_1(T)$ by the standard homomorphism, η, which in this case is an isomorphism. $H_1(S^3\text{-}k)$ is infinite cyclic so that the kernel of $\iota\eta$ is infinite cyclic. This kernel must in fact map isomorphically onto H_1 (component of S^3-T containing k) and as this latter group has a selected generator, (k is orientated) the kernel of $\iota\eta$ contains a generator mapping onto the homology class containing k. A simple closed curve on T may be selected [15] representing this generator. This oriented curve, ℓ, is called a <u>longitude</u>. The homotopy class containing ℓ generates a direct summand of $\pi_1(T)$. A generator of the other factor of $\pi_1(T) \approx Z \oplus Z$ may be selected so as to be represented by a simple closed curve, m, which is homotopic to 0 in the component of S^3-T, which contains k. We shall call m a <u>meridian</u>. We select that orientation of m so that m and k have linking number [58] +1. Fig. VII-1 may make this paragraph clearer.

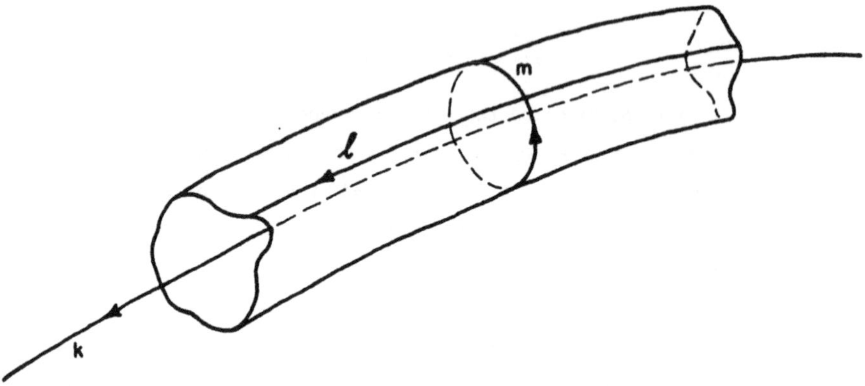

Fig. VII-1

We now have available two fixed conjugacy classes of elements in G, the longitudes, and the meridians. Much of the discussion in this chapter will utilize these elements.

In Chapter V the Smith Problem was mentioned briefly. (Can a knot be the fixed point set of a periodic homeomorphism of S^3?) Here again the existence or non-existence of automorphisms of various sorts throws some light on this problem. Recall that if the Smith problem has an affirmative answer, then for some n, K_n is isomorphic to a knot group H. Since K_n is a normal subgroup of finite index in another knot group G, there is an automorphism of H, which is periodic modulo an inner automorphism of H. We can assert still more: since k is left fixed this "periodic" automorphism leaves the set of longitudes fixed, and if the automorphism if of odd order, it leaves the set of meridians fixed. This connection having been made, we may in good conscience inquire into the properties of the automorphisms of knot groups.

§2. Outer Automorphisms

One of the earliest results concerning outer automorphisms of knot groups is due to O. Schreier. He proved in [55] that the group of a torus knot $\big((a, b ; a^p = b^q), (p, q) = 1\big)$ admits an outer automorphism, in fact he determined a set of generators and relations for the automorphism group of these groups. His result is expressed in Theorem 7.2.1.

THEOREM 7.2.1. (Schreier) The automorphism group of $|a, b ; a^p = b^q|$, $(p, q) = 1$ may be presented,
$(I, J, K ; I^p = J^q = K^2 = (KI)^2 = (KJ)^2 = 1)$
where I is conjugation by a, J is conjugation by b, and K maps a to a^{-1} and b to b^{-1}.

Later Magnus, in [36], determined the group of automorphisms of the knot 4_1.

It is not difficult to prove a theorem about the automorphisms of those knot groups whose commutator subgroup is finitely generated. We use φ to denote the automorphism of G' induced by a generator of G/G'.

THEOREM 7.2.2. If G' is finitely generated then the group of automorphisms G inducing the identity automorphism of G/G' is isomorphic to the subgroup of the group of automorphisms of the free group G', which commute with φ modulo an inner automorphism of G'.

PROOF. Recall that G may be defined by generators $t, x_1, x_2, \ldots, x_{2g}$, and by relations $tx_it^{-1} = \varphi(x_i)$ $i = 1, 2, \ldots, 2g$, where the x_i freely generate G', and φ is an automorphism of G'. If an automorphism, ψ, of G induces the identity on G/G', then t must be mapped into an element tc where $c \in G'$. Then we must have in G,

$$tc\psi(x_1)c^{-1}t^{-1} = \psi\varphi(x_1) \tag{1}$$

which implies

$$\varphi\left(c\psi(x_1)c^{-1}\right) = \psi\varphi(x_1) \tag{2}$$

so that

$$\varphi(c)\left(\varphi\psi(x_1)\right)\varphi(c^{-1}) = \psi\varphi(x) \tag{3}$$

and $\varphi\psi = \psi\varphi$ modulo an inner automorphism. Conversely, if there exists an automorphism ψ of G' satisfying $d\left(\varphi\psi(x_1)\right)d^{-1} = \psi\varphi(x)$ then $d = \varphi(c)$ for some c, and

$$t \to tc$$
$$x_1 \to \psi(x_1)$$

define an automorphism of G. This completes the proof.

It is not hard to see which automorphisms of G' extend to automorphisms of G inducing the non-trivial automorphism of G/G', namely instead of requiring

$$d\left(\varphi\psi(x_1)\right)d^{-1} = \psi\varphi(x_1) \tag{4}$$

we require

$$d\left(\varphi^{-1}\psi(x_1)\right)d^{-1} = \psi\varphi(x) \tag{4'}$$

that is, $\psi\varphi$ and $\varphi^{-1}\psi$ differ by an inner automorphism.

As a corollary to Theorem 7.2.2 and the remarks following, we shall show that the action of an automorphism on G is determined by its action on G', so we state:

COROLLARY 7.2.1. If G' is finitely generated, then the only automorphism of G that leaves each element of G' fixed is the identity.

PROOF. If ψ denotes the automorphism in question and if ψ is trivial on G/G' then we have by formula (4) above, since ψ is trivial on G',

$$d\,\varphi(x_i)d^{-1} = \psi\varphi(x_i) = \varphi(x_i) \quad \text{for all } i = 1, 2, \ldots, 2g,$$

for some $d \in G'$. But $\varphi(x_i)$ freely generate G' and G' is free of rank > 1 so d must equal 1. On the other hand, $d = \varphi(c)$, and $tc = \psi(t)$ so that $\psi(t) = t$. If ψ is non-trivial on G/G', then by (4') $d\,\varphi^{-1}(x_i)d^{-1} = \varphi(x_i)$ and we have

$$\varphi(d)x_i\varphi(d^{-1}) = \varphi^2(x_i) \quad . \tag{5}$$

Now consider this equation modulo G''. Denoting (as in Chapter IV) by M the matrix defining the action of φ on G'/G'', (5) taken modulo G'' implies that M^2 is the identity. But then every root of $\Delta_G(t)$ must be a square root of unity. This is impossible because G' is not perfect, so by Theorem 4.9.1 $\Delta_G(t) \not\equiv 1$, and the only other possibilities for factors of $\Delta_G(t)$, t^2+1, $t+1$, $t-1$, are not 1 when t is set equal to 1, thus $\Delta_G(1) \neq 1$. This completes the proof of the corollary.

An autohomeomorphism of S^3-k induces an automorphism of $\pi_1(S^3-k)$ which is defined up to an inner automorphism. Such an automorphism maps the group system onto itself. It is of interest to note that all the automorphisms of a knot group whose commutator subgroup is finitely generated which map the group systems into itself, do in fact, arise from atuohomeomorphisms of S^3. We state this formally in the following theorem.

THEOREM 7.2.3. If G' is finitely generated and ψ is an automorphism of G mapping the group system onto itself, then there exists an autohomeomorphism of S^3 which induces ψ on G.

PROOF. In [46], and Chapter X, it is proved that the complements of open tubular neighbourhoods of two knots with finitely generated commutator subgroups are homeomorphic if there exists an isomorphism between their groups mapping one group system onto the other, furthermore, this homeomorphism may be chosen so as to induce the given isomorphism. Specializing this homeomorphism proves the existence of an autohomeomorphism of S^3 minus an open tubular neighbourhood of the knot. This autohomeomorphism may be extended to all of S^3 because a meridian (which generates the kernel of the mapping from the fundamental group of the boundary of the tubular neighbourhood into the tubular neighbourhood) is mapped into a meridian (or its inverse) according to the part of the hypothesis relating to the group system.

This completes the proof of the theorem.

§3. Symmetries

For a long period of time it was unknown whether every knot could be inverted. That is: Is every knot equivalent to itself with opposite orientation? (Naturally, by an orientation preserving map of S^3.) Expressed algebraically this question may be weakened to the following: Does every knot group admit an automorphism mapping a longitude onto the inverse of a longitude, and a meridian onto the inverse of a meridian?

This long standing question was answered in the negative by Trotter [64] who proved that the groups of certain knots admit no such automorphism.

A similar but simpler question is whether a knot is equivalent to its mirror image. Such a knot is called amphicheiral. A corresponding (weaker) algebraic question is the following: Does every knot group admit an automorphism mapping a meridian onto the inverse of a meridian, and a longitude onto a longitude?

The answer to this question is "no." The counter example is the trefoil knot (3_1). The proof is provided in [14] and [15]. In the latter paper, Fox proves that the group of the trefoil times the trefoil's mirror image† is not equivalent to the product of the trefoil with itself. The method of proof used by Fox is to show that these two knots admit no isomorphism of their groups which preserves the group system. If an automorphism of the group of the trefoil existed, mapping a meridian onto the inverse of a meridian, and a longitude onto a longitude, then this automorphism could be followed by an inner automorphism so that the resulting composition would leave a longitude fixed. This would allow an isomorphism to be constructed between the group and the group system of the product of the trefoil knot with its mirror image, and the product of the trefoil knot with itself. Thus, Fox's proof also proves that _the group of the trefoil knot admits no automorphism mapping a meridian onto the inverse of a meridian and a longitude onto itself_.

While the title of this section is "Symmetries," up to now we have not required that the automorphisms which are induced by homeomorphisms of the residual space satisfy a prime requirement of symmetries, namely, that they be of finite order. As it is useful to speak of algebraic symmetries we make the following definition.

DEFINITION. An algebraic symmetry of a knot group is an automorphism of finite order.

Trotter investigated [65] the algebraic symmetries of knot groups whose commutator subgroups are free, and obtained the following theorem.

THEOREM 7.3.1. (Trotter [65] Lemma 3.3) If G' is finitely generated and G admits a symmetry of order q, then all the roots of $\Delta_G(t)$ are q^{th} roots of unity.

The symmetries of order 2 of a knot group seem to be to be of independent interest, in view of the various geometrical considerations leading to questions concerning them. In the next chapter there will be found a construction which is conceptually at least based upon the idea of extending the non-trivial symmetry of the inifinie cyclic group to an automorphism of a knot group. The end result is not a complete theory, but

† The product of two knots is defined, for example, in [56], [18] and Chapter VIII of this book.

rather some algebraic theorems which are hopefully capable of being connected with geometric ideas. The connection has not yet been made, so I hope it will be possible for someone to find out what (if any) connection may be made with a topological point of view.

CHAPTER VIII

A GROUP OF GROUPS

§1. Introduction

While there are a number of means available for composing one group with another, such as direct product, free product, wreath product, none of these seems to impose more structure on a set of groups than that of a semi-group with identity.

Similarly, topology abounds with methods for making one space from two; these include for example, $W \times V$, $W \cup V$, $W \vee V$, W^V etc. On the other hand these operations, again, rarely lead to anything with as much structure on it as a group. (Of course, two sparkling exceptions are the homotopy groups and their generalizations, and the cobordism groups. The latter in relative form, suggested the contents of this chapter.)

Partly in view of its rarity, partly for amusement, and finally, because it may be useful, we present in this chapter a means for constructing a group from a collection of groups or spaces. The collections are far from arbitrary, as will be seen from the axioms, while the rule of composition is suggested by the group theoretic operation of free product with amalgamation.

We shall specialize this general construction to the set of knot groups.

In the last section we prove that a group of knot groups of 2-spheres in S^4 is non-trivial.

§2. The Semi-Group of Knots

The physical operation of tying one knot on top of another may be formalized in the following manner. The two knots k_1, k_2, may be assummed to lie in disjoint closed 3-cells E_1, E_2. A closed arc of each α_1, α_2 may be assumed to lie in the interior of a disc D_1, D_2, on the boundary of E_1, E_2, while the rest of each knot lies in the interior of its containing 3-cell. If S^3 is given an orientation, then E_1, E_2, inherit orientations which induce orientations on D_1 and D_2. Give k_1, k_2 orientations so that α_1, α_2 are oriented. Now match (D_1, α_1) with (D_2, α_2) so that the orientations of D_1 as well as that of the α_1 are in disagreement. This process leads to an oriented knot lying in an oriented 3-cell, E, when the interior of the now common arc α_1, α_2, is removed and E is imbedded with its orientation preserved, in an oriented 3-sphere. It is not hard to see that the knot type we obtain, which we denote $k_1 \# k_2$, is independent of the arcs, 2-discs, and 3-cells chosen. This operation makes the set of knot types into a semi-group, and a construction of Fox and Milnor [23] may be used to define equivalence classes which form a group under the semi-group operation. (This construction, and the corresponding definitions may be done for n-knots in n+2-space [27], [28].)

The group of $k_1 \# k_2$ may be computed from the groups of k_1 and k_2 respectively, by application of the van Kampen Theorem to E_1, E_2, D_1 and D_2. If G_1 is the group of k_1, G_2 that of k_2, then there are elements (meridians) t_1, t_2 in G_1, G_2 so that

$$\pi_1(S^3 - (k_1 \# k_2)) = G_1 \underset{\{t:\}}{*} G_2$$

where t is identified with t_1 in G_1, and t_2 in G_2.

If we agree to select in each knot group a meridian which encircles the oriented knot by the left hand rule, then the semi-group of knots is mimicked by the semi-group of knot groups, with the operation in the latter being a free product with amalgamation which identifies <u>the</u>

homotopy class of the meridian in one group with <u>the</u> homotopy class of the meridian in the other.

Note: These semi-groups are not naturally isomorphic, as some distinct knots may have the same group and distinguished meridian, (e.g., granny and square).

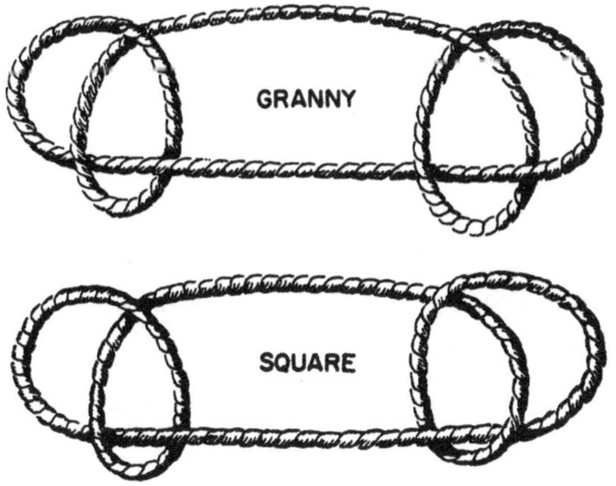

Fig. VIII-1

The semi-group of knot groups suggests a more general construction which is given in the next section.

§3. Some Axioms

We suppose we are given a set, A, of groups. There is a distinguished member of A, which we denote by G, and there is at least one monomorphism φ_H of G into each group H in A. We consider the set, $\underline{A}$, of <u>all</u> pairs (H, φ_H), and denote by Ψ the natural map from $\underline{A}$ to A.

These objects must satisfy the following axioms:

Ax. 1. There exists an involution $\iota: G \to G$.

Ax. 2. $(G, \text{identity}) \in \underline{A}$.

Ax. 3. If $(F, \varphi_F) \in \underline{A}$, then $(F, \varphi_F \iota) \in \underline{A}$.

Ax. 4. If (F, φ_F) and (H, φ_H) are in $\underline{A}$, then

$$(F \underset{\varphi_F(G) = \varphi_H(G)}{*} H, \varphi_F) \text{ is in } \underline{A} .$$

We remark that as F and H are canonically subgroups of the free product with amalgamation, the placement of G in $F \underset{G}{*} H$ is well defined by either φ_F or φ_H, and obviously imbedded in the same manner by either.

§4. A Binary Relation

Denote the binary operation on elements in $\underline{A}$ described in Axiom 4 by $(F, \varphi_F) \circ (H, \varphi_H)$.

Define a binary relation, r, between elements in $\underline{A}$ as follows:

$$(F, \varphi_F) r (H, \varphi_H), \text{ if and only if } \Psi((F, \varphi_F) \circ (H, \varphi_H \iota))$$

admits an involution which induces ι on the subgroup $\varphi_F(G)$.

CLAIM: r is symmetric.

PROOF. Consider $F \underset{\varphi_F(G) = \varphi_H(\iota(G))}{*} H = L$, and

$$L' = H \underset{\varphi_H(G) = \varphi_F(\iota(G))}{*} F .$$

Suppose $\iota(g') = g$, then substitution of $\iota(g')$ for g in the equation

$$\varphi_F(g) = \varphi_H(\iota(g)) \tag{1}$$

yields

$$\varphi_F(\iota(g')) = \varphi_H(\iota(\iota g')) = \varphi_H(g') . \tag{2}$$

But Equation (1) constitutes the amalgamating relations for L, while Equation (2) expresses the amalgamating relations for L', so that L and L' are isomorphic. This means that $(F, \varphi_F)\ r(H, \varphi_H)$ implies $(H, \varphi_H)\ r(F, \varphi_F)$.

CLAIM: r is reflexive.

PROOF. We must prove that $F \underset{\varphi_F(G) = \varphi_F(\iota(G))}{*} F = L$ admits an involution which induces ι on the subgroup $\varphi_F(G)$. So we seek an automorphism $e\colon L \to L$ such that

$$e(\varphi_F(G)) = \varphi_F(\iota(G)) \quad .$$

Let j_1, j_2 denote isomorphisms of F onto F_1, F_2 (Two distinct copies of F). Then $L \approx F_1 \underset{j_1\varphi_F(G) = j_2\varphi_F(\iota(G))}{*} F_2$. We now consider F_1, F_2 as imbedded in the natural way in L [33]. Consider the mappings $j_2j_1^{-1}$, and $j_1j_2^{-1}$. These mappings interchange F_1 and F_2. Consider an element $j_1(\varphi_F(g))$. $j_2j_1^{-1}$ maps this element onto $j_2\,\varphi_F(g)$, while $j_1j_2^{-1}$ maps $j_2(\varphi_F(g))$ onto $j_1(\varphi_F(g))$. Now, if $g = \iota g'$ then $g' = \iota g$, and $j_2j_1^{-1}\ j_1(\varphi_F(g)) = j_2(\varphi_F(g)) = j_2(\varphi_F(\iota(g'))$ while

$$j_1j_2^{-1}\ j_1(\varphi_F(g)) = j_1j_2^{-1}\ j_2\varphi_F(\iota(g)) = j_1\varphi_F(\iota g) = j_1\varphi_F(g'),$$

but since $j_1\varphi_F(g') = j_2\varphi_F(\iota g')$, $j_2j_1^{-1}$ and $j_1j_2^{-1}$ agree on $F_1 \cap F_2$ so that we may define a 1-1 mapping of the set $F_1 \cup F_2$ onto $F_1 \cup F_2$ by sending elements in F_1 onto elements in F_2 by means of $j_2j_1^{-1}$, and elements in F_2 onto elements of F_1 by means of $j_1j_2^{-1}$. Furthermore, since

$$(j_2j_1^{-1})j_1\varphi_F(g) = j_2\varphi_F(g) = j_2\varphi_F(\iota(g'))$$

and

$$(j_1j_2^{-1})j_2\varphi_F(\iota(g)) = j_1\varphi_F(\iota(g)) = j_1\varphi_F(g')$$

it follows that $j_2 j_1^{-1}$ and $j_1 j_2^{-1}$ may be extended to define an automorphism,† e, of L. When we note that $e^2 = \text{identity}$, and

$$e(j_1 \varphi_F(g)) = e j_2 \varphi_F(\iota(g)) = j_1 \varphi_F(\iota(g)) \quad ,$$

the claim is verified.

CLAIM: If $(F, \varphi_F)\ r(H, \varphi_F)$, and $(B, \varphi_B)\ r(C, \varphi_C)$ then

$$\left[(F, \varphi_F) \circ (B, \varphi_B)\right] r \left[(H, \varphi_H) \circ (C, \varphi_C)\right] \ .$$

PROOF. Let e denote the involution of $F \underset{G}{*} H$ which extends ι, and f the involution of $B \underset{G}{*} C$ which extends ι. Rewrite

$$\left(F \underset{\varphi_F(G) = \varphi_B(\iota(G))}{*} B\right) \underset{\varphi_F(G) = \varphi_H(\iota(G))}{*} \left(H \underset{\varphi_H(G) = \varphi_C(\iota(G))}{*} C\right) \quad \text{as}$$

(3)

$$\left(F \underset{\varphi_F(G) = \varphi_H(\iota(G))}{*} H\right) \underset{\varphi_F(G) = \varphi_B(\iota(G))}{*} \left(B \underset{\varphi_B(G) = \varphi_C(\iota(G))}{*} C\right) = N \ .$$

By hypothesis, each parenthesized factor of (3) admits an involution which induces ι on $\varphi_F(G) = \varphi_C(G)$, so that these involutions, e and f, may be extended to an involution, h of H, and h induces ι on $\varphi_F(G)$. This proves our claim.

Notice that $[(F, \varphi_F) \circ (F, \varphi_F \iota)]\ r\ (G, \text{identity})$, since $(F, \varphi_F)\ r(F, \varphi_F)$. This implies that given any element of $\underline{A}$, there exists another element in $\underline{A}$ such that the product of these two elements admits an involution extending ι.

---

† $j_2 j_1^{-1}$ and $j_1 j_2^{-1}$ define an automorphism of $F_1 * F_2$ which maps the kernel of the natural map of $F_1 * F_2$ onto L, onto itself.

If we now define $(F, \varphi_F) \equiv (H, \varphi_H)$ when there exist elements X_i and $\underline{A}$ such that $(F, \varphi_F)\ r\ X_1\ r\ X_2 \ldots r\ X_n\ r\ (H, \varphi_H)$, then $\equiv$ is an equivalence relation since r is reflexive and symmetric.

It follows easily from our last claim that the operation o is well-defined on equivalence classes. Clearly, the equivalence class containing (G, identity) acts as an identity. Associativity of the product operation on equivalence classes follows from the associativity of the product operation on elements. An inverse to each equivalence class exists by virtue of the remark following the verification of the last claim. Thus we have proved

> THEOREM 8.4.1. A set of groups satisfying axioms 1-4 may be made into a group with respect to the rule of composition o, by means of the equivalence relation generated by the binary relation r.

§5. Some Examples

With very little effort a version of Theorem 8.4.1 of the preceding section may be proved in the following context.

Let A be a set of disjoint topological spaces, G a fixed member of A, imbedded by φ_F in each space $F \in A$. Let $\underline{A}$ be the set of pairs (F, φ_F). Let ι be an autohomeomorphism of G, of period 2. Assume† $(G, id) \in \underline{A}$, and if $(F, \varphi_F) \in \underline{A}$, then $(F, \varphi_F\iota) \in \underline{A}$. Now define the operation o on pairs of elements of $\underline{A}$ as follows:

$$(F, \varphi_F) \circ (H, \varphi_H) = (F \cup H/\varphi_F(g) = \varphi_H(g), \varphi_F) \quad .$$

Define r as before, and prove without difficulty the claims necessary to verify Theorem 8.4.1 in the above context.

† We denote the identity map by id.

As a simple example of this group of equivalence classes of topological spaces we may take as A the "whisker" spaces shown below

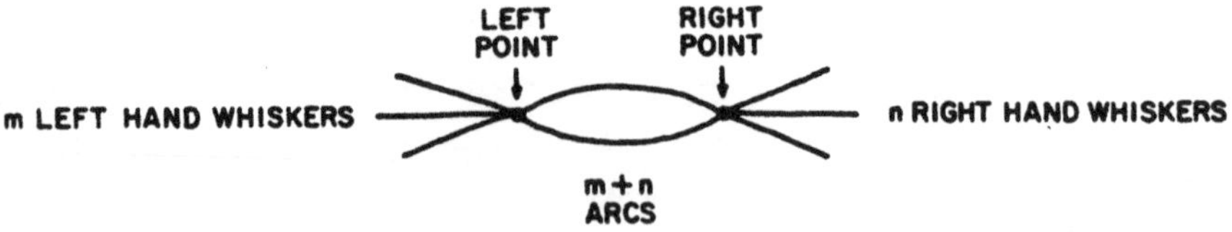

$m, n \geq 0.$

Take G to be the two point space $(m = n = 0)$, ι to be the interchange of these two points, and as imbeddings the maps putting these points (in either order) at the place indicated by the arrows in the above figure. It can be seen that two whiskers (F, φ_F), (H, φ_H) are equivalent if and only if the sum of the whiskers emanating from the image of point 1 in F, and point 2 in H is equal to the sum of the whiskers emanating from the image of point 2 in F and point 1 in H. Thus,

It thus becomes apparent that the group we have constructed is isomorphic to the integers. The isomorphism may be realized by mapping a pair (F, φ_F) onto the integer (number of whiskers emanating from the image of 1) minus (number of whiskers emanating from the image of 2).

Another topological adaptation of the construction is to take for the operation $\circ$, the cartesian product of spaces with base points, with one subspace identified with another. More precisely, if $\left((X, x_0), \varphi_x\right)$ and $\left((Y, y_0), \varphi_y\right)$ are two spaces with basepoints and φ_x, φ_y imbed a space (G, g_0) in each, then define

$$\left((X, x_0), \varphi_x\right) \circ \left((Y, y_0), \varphi_y\right) = \left((Z, z_0), \bar{\varphi}_x\right)$$

where

$$Z = X \times Y/(\varphi_X(g),\ y_0) = (x_0,\ \varphi_Y(g)) \ ,$$
$$z_0 = (x_0,\ y_0) = (\varphi_X(g_0),\ \varphi_Y(g_0)) \ ,$$
$$\bar{\varphi}_X\ (g) = (\varphi_X(g),\ y_0) = (x_0,\ \varphi_Y(g)) = \bar{\varphi}_Y(g) \ .$$

Proceeding as before, a set of pairs, consisting of a space with a base point, and an imbedding of a fixed space with a basepoint, may be made into a group using the operation described above, providing the set satisfies Axioms 1-4 interpreted in the topological context.

Although these examples are topological, Axiom 4 can also be modified in a purely algebraic way. Instead of free product with amalgamation, we may use direct product with amalgamation. With this modification and some simple condition on the groups in question, such as all are abelian, all the proofs will still go through, and a group may again be constructed.

Before passing on to the consideration of knot groups, a little discussion of the previous constructions is in order.

§6. Amalgamations

It is clear from the large number of interpretations of the axioms, and definition of o, which still lead to a group, that there should be a more general setting for §3 and §4.

While one may operate in a more general setting algebraically, (wherever products with amalgamation may make sense, e.g., for lattices, or direct products with amalgamation of abelian groups) it is somewhat more difficult to make a general definition of o which will include the topological interpretations as well. Furthermore, it is probably also possible to make the construction of "product with amalgamation" on projective planes, algebras, modules, semi-groups, lattices etc. If one can in fact, make the "proper" definitions then these can be used to redo §3 and §4 in the appropriate context.

For the present, we are content to revise §3 and §4 as necessary to cope with each particular set of structures, and "product with amalgamation." In the Appendix Prof. Eilenberg demonstrates the proper general setting for all these constructions.

§7. Knot Groups

Let us take for a set, A, the set of all knot groups, one for each oriented knot type. As in Chapter VII, Fig. VII-1, in each knot group certain elements (meridians) may be represented by a path α from a basepoint to a point near the knot, a little loop winding once positively about the knot, and the path α^{-1}.

Let us take for G, the integers. (The group of the trivial knot.)

Let us take for the elements of $\underline{A}$, pairs consisting of a knot group F, and all imbeddings of G in F which map a generator of G onto a meridian. These are imbeddings because F is without elements of finite order. [51]

It is easy to see that this set $\underline{A}$ satisfies the axioms, with ι interchanging +1 and -1.

The operation $\circ$ is induced by the operation # on knot types.

It is easy to see that if (F, φ_F), (F, ψ_F) are two pairs arising from a knot group F, and mappings of +1 onto two distinct meridians with the same linking number with the knot, k, then $(F, \varphi_F)\ r\ (F, \psi_F)$ so that (F, φ_F) is equivalent to (F, ψ_F). This may be proved as follows:

$$\Psi\Big((F, \varphi_F) \circ (F, \psi_F \iota)\Big) = L$$

is isomorphic to

$$\Psi\Big((F, \varphi_F) \circ (F, \varphi_F \iota)\Big) = L'$$

by a mapping, say ρ. A meridian m' is conjugate (as are all meridians) to $\rho(m)$ where m is a meridian in L, since both L and L' arise from the knot type $k \# k$. L, of course, admits an involution, e, which induces ι. If $m' = w\,\rho(m)w^{-1}$, then $\ell \to \left(\rho\left(e\left(\rho^{-1}(\ell^{w})\right)\right)\right)^{w^{-1}}$ defines an involution on L' which induces a non-trivial involution on the infinite cyclic group generated by $w(\rho(m))w^{-1}$. Thus, any meridian in L' may be mapped onto its inverse by an involution of L'. It follows then that (F, φ_F) is equivalent to (F, ψ_F).

While this is heartening it is far from satisfactory. Unfortunately, the group of knot groups may be trivial, although this is highly unlikely.

(It is not difficult to prove that the pair $\left((a, b\colon\ a^2 = b^3),\ \varphi\colon\ 1 \to ab^{-1}\right)$ admits no involution sending ab^{-1} onto ba^{-1}. While this proves that this pair is not r-related to the integers it does not prove that this pair is a non-trivial element in the group of knot groups.)

If we consider the set of knot groups and meridians of knotted 2-spheres in the 4-sphere which have Alexander Polynomials [32], the situation is somewhat better.

An automorphism of a knot group mapping a meridian onto its inverse induces the substitution $t \to t^{-1}$ in a matrix presenting the abelianized commutator subgroup as a module over the integral group ring of the knot group made abelian. (Chapter IV.) This implies that the Alexander Polynomial of a knot group admitting such an automorphism is symmetric.

Now suppose (F_0, φ_0) is a pair in the equivalence class of the identity in the group of knot groups of 2-spheres in the 4-sphere. We must have $(F_0, \varphi_0)\ r\ (F_1, \varphi_1)\ r\ \ldots\ r\ (F_n, \varphi_n)\ r\ (G, id)$. Since $(F_n, \varphi_n)\ r\ (F_{n-1}, \varphi_{n-1})$, and $(F_n, \varphi_n)\ r\ (G, id)$, F_n and

$$F_n \underset{\varphi_n(G) = \overset{*}{\varphi}_{n-1}(\iota(G)}{} F_{n-1} = L_n$$

must admit involutions inducing ι. By the argument above, the Alexander Polynomials of F_n, and L_n are symmetric. Since the polynomial of L_n is the product of the polynomials of F_{n-1} and F_n, and since the product of a non-symmetric polynomial and a symmetric polynomial cannot be symmetric, it follows that the polynomial of F_{n-1} must be symmetric. Moving up the line by the same argument, we may conclude that the polynomial of F_{n-2} is symmetric. Proceeding in this fashion, we may conclude;

THEOREM 8.7.1. If (F, φ_F) is equivalent to (G, id), then the Alexander Polynomial of F is symmetric.

From this the following two corollaries are immediate.

COROLLARY 8.7.1. If (F, φ_F) is equivalent to (H, φ_H) then $\Delta_F(t)\, \Delta_H(t^{-1})$ is symmetric.

COROLLARY 8.7.2. The group of knot groups of 2-spheres in the 4-sphere which have Alexander Polynomials is non-trivial.

PROOF. There exists $S^2 \subset S^4$ with a non-symmetric polynomial [32].

CHAPTER IX

THE CHARACTERIZATION PROBLEM

§1. Introduction

Probably the most elusive aspect of the study of knot groups is just the capture of those algebraic properties which characterize these groups. This chapter contains essentially all that has been published on this question. These results fall naturally into three categories, necessary and sufficient conditions, sufficient conditions, and necessary conditions, that a group be a knot group. The organization of this chapter follows this trichotomy.

§2. Necessary and Sufficient Conditions

The following disturbing, but not entirely discouraging result of John Stallings [60] gives one pause.

> THEOREM 9.2.1. (Stallings) There exists no algorithm for deciding whether or not an arbitrary finitely presented group is the group of a knot.

This theorem indicates the difficulty of the problem of characterizing the knot groups. It limits the kind of characterization one may hope for. It should not have the effect of discouraging attempts to provide methods or theorems useful in answering the basic question: What is a knot group?

The following theorem of Artin [3] provides an interesting example of a result which is available in spite of Theorem 9.2.1. The proof of Artin's theorem depends on the ideas of a "Braid," and while they are appealing it would take us a bit far afield to reproduce enough of them to prove this theorem.

THEOREM 9.2.2. (Artin) A group is a knot group if and only if it has a presentation of the following form:

$$(t_1, t_2, \ldots, t_n; \ T_i t_i T_i^{-1} = t_{i+1}, T_n t_n T_n^{-1} = t_1)$$

$$i = 1, 2, \ldots, n-1$$

where T_i are words in the t_i satisfying

$$\prod_{i=1}^{n} T_i t_i T_i = \left(\prod_{i=2}^{n} t_i \right) t_1$$

in the free group generated by $t_1, t_2, \ldots, t_n$.

The frequency with which we have dealt with knot groups whose commutator subgroup is free makes it desirable to have a characterization of these groups. Unfortunately, half the theorem we present here depends on the (as yet) unresolved status of the Poincaré conjecture.†

Let us suppose from now on that G is a group which extends the free group F_{2n} by the infinite cyclic group Z. We denote by t, an element of G which maps onto a generator of Z, and by 0 is meant the trivial group.

THEOREM 9.2.3. If
(1) $F_{2n} = [G, G]$,
(2) $G/(t) = 0$,

(3) $\left[t, \prod_{i=1}^{n} [a_i, b_i]\right] = 1$ in G,

† A simply connected closed 3-manifold is homeomorphic to the 3-sphere.

where a_i, b_i, $i = 1, 2, \ldots, n$ generate F_{2n}, then, if the Poincaré conjecture is true, there exists a tame knot k in S^3, such that $\pi_1(S^3 - k) \approx G$. Conversely, $\pi_1(S^3 - k)$ satisfies (1), (2), and (3), if $[G, G]$ is finitely generated.

PROOF. We shall construct over the circle, a fiber space whose fundamental group is G. The total space will be the complement of an open tubular neighbourhood of a knot in a simply connected 3-manifold.

Let S denote an orientable 2-manifold of genus n, with boundary a single curve ℓ.

Now G is completely described when the automorphism of F_{2n} induced by t is described. Denote this automorphism by φ. We shall consider φ as acting on $\pi_1(S)$; then condition (3) implies that φ leaves a class of loops containing ℓ, with some orientation, fixed. By doubling S along ℓ, and applying Nielsen's Theorem in [48] we see that φ is induced by some autohomeomorphism Φ of S.†

Now by matching the two ends of $I \times S$ according to Φ, one obtains a 3-manifold M (with boundary T a torus) whose fundamental group is G. Furthermore, by hypothesis, adjoining the relation $t = 1$ to G destroys G completely. Thus, by matching the boundary of a solid torus $D^2 \times S^1$ with T so as to kill t in G, one may obtain a simply connected manifold, such that M is the complement of a tubular neighbourhood of a knot, (an interior point of D^2) $\times S^1$, and half the theorem is proved.

The necessity of condition (2) follows immediately from the fact that S^3 is simply connected and does not depend upon the structure of $[G, G]$.

---

† Nielsen's Theorem states that any automorphism of the fundamental group of a closed orientable 2-manifold is induced by an autohomeomorphism of the 2-manifold.

The conditions (1) and (3) are implied by the finite generation of $[G, G]$ and follow from Theorem 4.5.1. One need only note that when $[G, G]$ is finitely generated, $F_{2n} \approx \pi_1(S)$ (for S an orientable surface of minimal genus spanning k) is imbedded in G as $[G, G]$, and the element of $\pi_1(S)$ corresponding to the class of ∂S (base point on ∂S) satisfies (3).

It should be noted that condition (3) is satisfied if conjugation by t maps

$$\prod_{i=1}^{n} [a_i, b_i] \quad \text{onto} \quad w^{-1} \prod_{i=1}^{n} [a_i, b_i]\, w$$

where $w \in F_{2n}$; for in this case, we may take for t (which was picked as a preimage of a generator of G/G') the element wt which obviously commutes with $\Pi_{i=1}^{n} [a_i, b_i]$.

§3. Sufficient Conditions

The following theorem [74] is of some interest.

THEOREM 9.3.1. (H. Zieschang) Suppose $G \approx F_n \underset{F_{2n-1}}{*} F_n'$, where F_j', F_j are free of rank j. Suppose $F_{2n-1} \subset F_n$ is generated by $a_2, \ldots, a_{2n}$, and $F_{2n} \subset F_n'$ is generated by $b_2, \ldots, b_{2n}$ where $a_i = b_i$ in G. Suppose that there exists $a_1 \in F_n$, $b_1 \in F_n'$ such that

$$\prod_{i=1}^{n} [b_{2i-1}, b_{2i}] = 1 \quad \text{in} \quad F_n' .$$

Suppose $G/(a_1 b_1^{-1}) = 0$. Then if the Poincaré conjecture is true in dimension 3, G is a knot group.

§4. Necessary Conditions

1. We have seen earlier (Chapter V) that a knot group G is "without" homology with simple coefficients in dimensions greater than one. This is quite a stringent condition, however it is most certainly true for a more general class of groups, as is almost any other particular property of knot groups.

2. Papakyriakopoulous showed [52] that a knot group which is not cyclic has 1 end. For a discussion of ends see [25] or [29].

3. As was mentioned in the introduction, G may be generated by a set of conjugates. This also follows from the presentation derived in Chapter III.

4. A knot group, by Chapter III and a simple computation of the Euler characteristic, has a presentation in which the number of generators exceeds the number of relations by exactly 1. Since the group made abelian is infinite cyclic, this is the maximum excess of generators over relations in any presentation, hence the knot group has deficiency 1.

5. Torres and Fox [62] discovered that a knot group has a presentation which admits a duality of a certain sort. This is expressed in the theorem below. A good proof of this theorem may be found in [12].

> THEOREM 9.4.1. (Torres and Fox) A knot group G has a presentation $(x_1, x_2, \ldots, x_n; r_1, r_2, \ldots, r_n)$ and a presentation $(y_1, y_2, \ldots, y_n; s_1, s_2, \ldots, s_n)$ satisfying the following equations for all i, $j = 1, \ldots, n$.

$$\Psi\Phi_x(x_i) = \Psi\Phi_y(y_i^{-1})$$

$$\Psi\Phi_x\left[\frac{\partial r_j}{\partial x_i}(x_i - 1)\right] = \Psi\Phi_y\left[\frac{\partial s_i}{\partial y_j}(y_j - 1)\right]$$

Here $\Phi_x(\Phi_y)$ maps the integral free group ring generated by $x_i(y_i)$ onto the integral group ring of G, and Ψ maps (in the obvious way) the integral group ring of G onto the integral group ring of G/G'.

This theorem is extremely important, for it implies the symmetry of the Alexander Polynomial of G, and serves to rule out as knot groups many candidates for this honor. The proof of this theorem is quite geometric, and we do not present it here because of considerations of space.

CHAPTER X

THE STRENGTH OF THE GROUP

§1. Introduction

Having spent our time up to now on an investigation of the knot groups themselves, it is appropriate that we complete this study with a chapter whose chief consideration is the extent to which the knot group determines the knot or the topological type of the complement of the knot.

§2. Homotopy Type

The proof by Papakyriakopoulous of the asphericity of the complement of a knot [51] "solves" the problem of the homotopy classification of the complements of knots. Since $\pi_n(S^3 - k) = 0$ for $n > 1$ and any k, any map of the 2-skeleton of the complement of a small tubular neighbourhood of a knot k into the 1-skeleton of the complement of a small tubular neighbourhood of another knot k' may be extended to 3-skeleton. It follows then that an isomorphism and its inverse between two knot groups may be realized by continuous mappings between the complementary domains of the two knots. As these continuous maps induce isomorphisms between the homotopy groups of the spaces in question, these mappings induce a homotopy equivalence [70]. In summary, $S^3 - k$ and $S^3 - k'$ are of the same homotopy type if and only if $\pi_1(S^3 - k)$ and $\pi_1(S^3 - k')$ are isomorphic.

§3. The Topological Type of the Complement of a Knot

The topological inequivalence of the complement of the square knot and the granny knot was proved in [14] by Dehn and again by Fox in [15].

The groups of these two knots are however, isomorphic. This fact indicated the need for a stronger algebraic invariant than the group alone. We have already seen a stronger invariant in Chapter VII where the group system was defined.

It has long been believed that the group system determines the topological type of the complement. This as yet has not been verified, although the following theorem may be proved.

THEOREM 10.3.1. If the commutator subgroup of a knot group is finitely generated, then the group system determines the topological type of the complement.

PROOF. Let $M(k)$ denote the closure of the complement of a regular neighbourhood of k. Then it is well known that the topological type of $M(k)$ depends only on the type of k. Now suppose k_1, k_2, two tame knots in S^3 are given, and there exists an isomorphism p, mapping $\pi_1(M(k_1))$ onto $\pi_1(M(k_2))$ and mapping the conjugacy class of $\pi_1(\dot{M}(k_1))$ onto that of $\pi_1(\dot{M}(k_2))$. We may assume without loss of generality that the base points, x_1, x_2, for $\pi_1(M(k_1))$, $\pi_1(M(k_2))$ are located on the boundary in each case; also by a suitable automorphism of $\pi_1(M(k_2))$, p may be assumed to map $\pi_1(\dot{M}(k_1)) \subset \pi_1(M(k_1))$ onto $\pi_1(M(k_2)) \subset \pi_1(M(k_2))$.

Let m_1, ℓ_1 denote a meridian and longitude of $\dot{M}(k_1)$, let m_2, ℓ_2 denote a meridian and longitude of $M(k_2)$. Then

$$p[m_1] = \pm(m_2 + r[\ell_2]),$$

and

$$p[\ell_1] = \pm [\ell_2] \quad .$$

It is easy to see that there exists a homeomorphism f from $\dot{M}(k_1)$ to $\dot{M}(k_2)$ such that $f^{\#} = p$.

Let S_1', S_2' denote surfaces of minimal genus spanning k_1, k_2 respectively and chosen so that $\dot{M}(k_i) \cap S_i' = \ell_i$. According to Theorem 4.1, and our assumption of the finite generation of $[\pi_1, \pi_1]$, these surfaces are of the same genus. $M(k_i) \cap S_i'$ will be denoted S_i.

We now wish to extend f to a homeomorphism f_2 from $M(k_1)$ to $M(k_2)$: we first extend f to a homeomorphism f_1 from $\dot{M}(k_1) \cup S_1$ to $\dot{M}(k_2) \cup S_2$, such that $(f_1 | S_1)^{\#} = p|[\pi_1, \pi_1]$, where this last equation has meaning by virtue of Theorem 4.5.1, where it is proven that $\pi_1(S_1)$ generates $[\pi_1, \pi_1]$ if $[\pi_1, \pi_1]$ is finitely generated. More formally, f_1 will be constructed so that the following diagram is commutative:

$$\begin{array}{ccc} [\pi_1, \pi_1] & \xrightarrow{\ p|[\pi_1, \pi_1]\ } & p[\pi_1, \pi_1] \\ \uparrow i & & \uparrow j \\ \pi_1(S_1) & \xrightarrow{\ (f_1|S_1)^{\#}\ } & \pi_1(S_2) \end{array}$$

Here i and j are isomorphisms by Lemma 4.4.2.

Let ϕ denote any homeomorphism of S_1 to S_2 which agrees with f on $S_1 \cap \dot{M}(k_1) = \ell$. (Such exists since S_1 and S_2 are of the same genus.) Now $\phi^{\#} \circ i^{-1} \circ p^{-1} \circ j$ defines an automorphism of $\pi_1(S_2)$ which leaves $\pi_1(S_2 \cap \dot{M}(k_2)) \subset \pi_1(S_2)$ fixed, so by attaching a disc, D, to S_2 along $\dot{S}_2$ and applying Nielsen's theorem [48] to the induced automorphism on $S_2 \cup D$, it is easily seen that there exists a homeomorphism $h\colon S_2 \cup D \to S_2 \cup D$ leaving D fixed such that $(h|S_2)^{\#} = \phi^{\#} \circ i^{-1} \circ p^{-1} \circ j$. Now define $f_1|S_1 = h^{-1} \circ \phi$. Thus $(f_1|S_1)^{\#} = j^{-1} \circ p \circ i$. Since $h^{-1} \circ \phi = f$ on $S_1 \cap \dot{M}(k_1)$, $h^{-1} \circ \phi$ may be extended to a homeomorphism f_1 on $\dot{M}(k_1) \cup S_1$, by defining f_1 on $\dot{M}(k_1)$ to be equal to f, and

f_1 on S_1 to equal $h^{-1} \circ \phi$. It is easy to see that f_1 may be extended to a homeomorphism f_2 from a nice small neighbourhood N of $\dot{M}(k_1) \cup S_1$ to a nice small neighbourhood of $\dot{M}(k_2) \cup S_2$. Now $\dot{N} - \dot{M}(k_1)$ is a 2-manifold, and clearly the component of $S^3 - (\dot{N} - \dot{M}(k_1))$ which does not contain k_1 is homeomorphic to $S^3 - S_1$, but by Theorem 4.5.1, $\pi_1(S^3 - S_1)$ is free, and an easy application of the Loop Theorem [50] and the Dehn Lemma [51] shows that the component of $S^3 - (\dot{N} - \dot{M}(k_1))$ not containing k_1 is a solid torus.

The problem has now been reduced to extending f_2, a homeomorphism defined on the boundary of a solid torus to a homeomorphism of the solid torus, more explicitly, f_2 maps N onto a homeomorphic neighbourhood of $\dot{M}(k_2) \cup S_2$, and the complements of N, and its homeomorph are solid tori, whose boundaries T_1, T_2 are homeomorphic by the map f_2.

By a well-known principle (see for example, [42]) a homeomorphism H from the boundary of one solid torus R_1 to another, R_2, may be extended to the interior if and only if $H^{\#}$ maps the kernel of the map $\iota^{\#}: \pi_1(\dot{R}_1) \to \pi_1(R_1)$ onto the kernel of the map $\eta^{\#}: \pi_1(\dot{R}_2) \to \pi_1(R_2)$.†

First we show that f_2 maps loops (on T_1) homotopic to 0 in $S^3 - N$ into loops (on T_2) homotopic to 0 in $S^3 - f_2(N)$.

Suppose α is a loop on T_1 not homotopic to 0 on T_1, but homotopic to 0 in $S^3 - N$. If $f_2(\alpha)$ is not homotopic to 0 in $S^3 - f_2(N)$, $f_2(\alpha)$ determines a non-trivial conjugacy class of $p([\pi_1, \pi_1])$ by Lemma 4.4.2. But by assumption, α is homotopic to 0 in $S^3 - N$, hence it determines the trivial conjugacy class in $[\pi, \pi]$ and so the trivial conjugacy class in $p([\pi, \pi])$. Arguing similarly with f_2^{-1}, allows the above principle to be applied, so that f_2 may be extended to f_3, which maps $M(k_1)$ homeomorphically onto $M(k_2)$.

† This follows easily from Dehn's Lemma.

§4. Knot Type

The passage from the topological type of $M(k)$, the complement of a small tube about the knot, to the knot type itself depends solely on the placement of a meridian in $\pi_1(M(k))$.

For example, if $M(k)$ and $M(k')$ are topologically equivalent by a map which sends a meridian onto a meridian, then this topological equivalence may be extended to a map from S^3 onto S^3 which sends k onto k'. (This follows immediately from the principle used in the proof of Theorem 10.3.1.)

We state the combination of this observation and Theorem 10.3.1 as Corollary 10.4.1.

> COROLLARY 10.4.1. If two knot groups are isomorphic by a map which sends a meridian onto a meridian, and the group system of one onto the group system of the other, then the knots are equivalent.

As far as orientation is concerned one can require the isomorphism to preserve the selection of oriented longitude and meridian if the autohomeomorphism is to preserve orientation of S^3 and the knot.

There are a few isolated results which should be mentioned in connection with these matters.

H. Gluck has proved [24] that any knot whose complement is homeomorphic to the complement of a torus knot is equivalent to that torus knot. Thus, Theorem 10.3.1 and Gluck's result prove that a torus knot is determined by its group system.

H. Zieschang has proved [74] that if the groups of two knots k, k' admit structures and isomorphisms of the following sort:

$$
\begin{array}{ccc}
(W_1, \ldots, W_n :) & * & (X_1, \ldots, X_n :) \\
\cup & & \cup \\
\left(H_1, \ldots H_{2n} : \prod_{i=1}^{n} [H_{2i-1}, H_{2i}]\right) & & \left(K_1, \ldots K_{2n} ; \prod_{i=1}^{n} [K_{2i-1}, K_{2i}]\right) \\
\cup & & \cup \\
& (H_2, \ldots H_{2n} :) = (K_2, \ldots, K_{2n}) &
\end{array}
$$

$$
\Big\downarrow\ \varphi : \begin{array}{l} W_i \to Y_i \\ X_i \to Z_i \\ H_i \to L_i \\ K_i \to M_i \end{array}
$$

$$
\begin{array}{ccc}
(Y_1, \ldots, Y_n :) & * & (Z_1, \ldots, Z_n :) \\
\cup & & \cup \\
\left(L_1, \ldots, L_{2n} : \prod_{i=1}^{n} [L_{2i-1}, L_{2i}]\right) & & \left(M_1, \ldots, M_{2n} ; \prod_{i=1}^{n} [M_{2i-1}, M_{2i}]\right) \\
\cup & & \cup \\
& (L_2, \ldots, L_{2n} :) = (M_2, \ldots, M_{2n} :) &
\end{array}
$$

Then k and k' are equivalent.

CHAPTER XI

PROBLEMS

§1. Introduction

The problems included in this chapter are of a research nature. They obviously reflect my own interests. I have tried however, to include such questions as have bearing on aspects of knot theory other than the group theoretic one. The reader may decide for himself those which are of a whimsical nature.

§2. Problems

A. Characterize the knot groups among the finitely presented groups with commutator quotient group Z.

B. Verify or disprove the following conjecture: Every knot group G is a non-trivial free product with amalgamation, the amalgamating subgroup being free.
(Background for this conjecture will be found in [43].)

C. Characterize algebraically the peripheral subgroups of a knot group: e.g., can the peripheral subgroups be located solely from a presentation of a knot group?

D. In connection with C: Is a peripheral subgroup a maximal abelian subgroup?

E. Is every knot group residually finite?

F. When is the kernel of the homomorphism from a knot group to Z_n isomorphic to a knot group? (Chapter III and [16].)

G. If a knot group has a center is it the group of a torus knot? (Chapter III and [44].) Murasugi had also asked this question.

H. Suppose the fundamental group of a closed 3-manifold, M, is a free product with amalgamation of two knot groups, $\pi_1(S^3 - k_1)$, $\pi_1(S^3 - k_2)$, with the amalgamating subgroup a peripheral subgroup of each knot group. Is this 3-manifold $M = M(k_1) \cup M(k_2)$ where $\partial M(k_1)$ and $\partial M(k_2)$ are matched by a homeomorphism. ($M(k_i)$ denotes the complement in S^3 of a small tube about k_i.)
(A related problem, incorrectly stated however, appeared in [17].

I. Is the group of knot groups (Chapter IX) non-trivial?

J. Suppose G' is finitely generated (so that G' is free by Theorem 4.5.1). G is described by an automorphism φ of G' (Chapter X). Can, $\varphi = \psi^n$ for $n > 1$ and ψ an automorphism of the free group G'? Giffen proved in his thesis that there exist knots for which $\varphi = \psi^n$ modulo an inner automorphism. (This is related to the Smith conjecture (Chapter III) as it applies to knots with G' finitely generated.)

K. Can Case B) of Theorem 4.5.1 actually occur?†

L. Is the decomposition in Case A) of Theorem 4.5.1 unique in the sense that the rank of the amalgamating subgroup is always twice the genus?

† See the first footnote to Chapter IV §5.

M. Does there exist a knot group with a non-trivial symmetry leaving a peripheral subgroup (Chapter VII) element-wise fixed? (This is sort of an algebraic version of the Smith problem.)

N. Can a knot group be ordered? (This is easy for knot groups with G' free.)

O. What can be said of the Frattini subgroup of a knot group?

P. Is there a simple condition on an automorphism φ of a free group F, of rank $2g$ which implies that $\Pi_{i=1}^{g} [a_i, b_i]$ is left fixed by φ? Here a_i, b_i are a set of free generators of F. The interest in this question of course, stems from the hypothesis of Theorem 9.2.3.

Q. Does the commutator subgroup of every knot group have cohomological dimension 1? (This would provide an example of a group with geometric dimension 2, category 2, and cohomology dimension 1. Perhaps in any case a perfect commutator subgroup (see Chapter IV) of a knot group is an example of such a group).

R. Does there exist a useful algebraic theory suggested by the Morse theory in dimension three? This might involve a generalization of Stallings' theorem [59].

S. Can a knot group contain an element $g \neq 1$ such that the equation $x^n = g$ has solutions for arbitrarily large values of n?

T. Every knot group contains the group $(a, b; [a, b])$. This subgroup may be obtained from the natural inclusion of the fundamental group of a non-singular torus in the knot group. Suppose a knot contains the group of a closed surface of genus g. Does there exist a non-singular colsed surface of genus g whose fundamental group is injected monomorphic ally into the knot group by the natural inclusion?

U. Suppose H is a group satisfying

a) $H/H' \approx Z_n$;

b) H may be finitely presented with at least as many relations as generators.

Is H a homomorph of a knot group G ?

V. An arbitrary knot group G must contain

a) A free group of any rank;

b) A free abelian group of rank 2.

Must G contain any other groups?

(By Theorem 4.5.1, G contains a free group of rank 2.)

W. Can the word problem be solved in a knot group? (Perhaps this follows from Theorem 4.5.1 or a positive solution to Problem B.)

X. Can one decide algebraically if a knot group is cyclic?

Y. Can one select geometrically significant representatives from each conjugacy class in a knot group?

Z. Can the crookedness of a knot type [49] be algebraically determined?

APPENDIX

by

S. Eilenberg

Let A be a category (which for the sake of simplicity will be assumed to be small, i.e., the objects of A will be assumed to be a set). Let G be a fixed object of A. Two morphisms $\varphi\colon G \to F$, $\varphi'\colon G \to F'$ in A are said to be isomorphic if there exists an isomorphism $\Phi\colon F \to F'$ such that $\Phi\varphi = \varphi'$. For any $\varphi\colon G \to F$ we denote by $[\varphi]$ the isomorphism class defined by φ. Let Φ be the set of all such isomorphism classes.

We define in Φ, the structure of a commutative monoid as follows. Given

$$\varphi_i\colon G \to F_i \qquad i = 1, 2$$

consider a pushout diagram

$$\begin{array}{ccc} G & \xrightarrow{\varphi_1} & F_1 \\ {\scriptstyle\varphi_2}\downarrow & & \downarrow{\scriptstyle\alpha_1} \\ F_2 & \xrightarrow[\alpha_2]{} & F \end{array}$$

This means that the diagram above is commutative and that given any commutative diagram

$$\begin{array}{ccc} G & \xrightarrow{\varphi_1} & F_1 \\ {\scriptstyle\varphi_2}\downarrow & & \downarrow{\scriptstyle\alpha_1'} \\ F_2 & \xrightarrow[\alpha_2']{} & F' \end{array}$$

there exists a unique $\gamma\colon F \to F'$ such that $\alpha_i' = \gamma\alpha_i$, $i = 1, 2$.

The existence of such pushouts for any given φ_1, φ_2 is postulated on the category A. Then define

$$[\varphi_1] + [\varphi_2] = [\alpha_1\varphi_1] = [\alpha_2\varphi_2] \quad .$$

The verification that Φ is a commutive monoid follows trivially from formal properties of pushouts. The zero in Φ is the class $[1_G]$.

Now assume that a morphism ι: $G \to G$ such that $\iota\iota = 1_G$ is given. A class $[\varphi] \in \Phi$ represented by φ: $G \to F$ will be called negligible if there exists a morphism ι': $F \to F$ such that

$$\iota'\iota' = 1_F \quad , \quad \varphi\iota = \iota'\varphi \quad .$$

The existence of such a ι' is clearly independent of the choice of the representative φ.

Let Ψ be the subset of Φ consisting of the negligible classes. One verifies easily that Ψ is a submonoid of Φ. Define the quotient monoid Φ/Ψ by defining an equivalence relation

$$[\varphi_1] \sim [\varphi_2]$$

provided

$$[\varphi_1] + [\psi_1] = [\varphi_2] + [\psi_2]$$

for some elements $[\psi_1], [\psi_2] \in \Psi$.

The general form of Theorem 8.4.1 may now be stated as

Theorem A. Φ/Ψ is a group.

Proof. Only the existence of an inverse needs to be shown. Given $[\varphi] \in \Phi$, we have $[\varphi\iota] \in \Phi$ and we assert that $[\psi] = [\varphi] + [\varphi\iota] \in \Psi$. By definition ψ is the diagonal in a pushout diagram

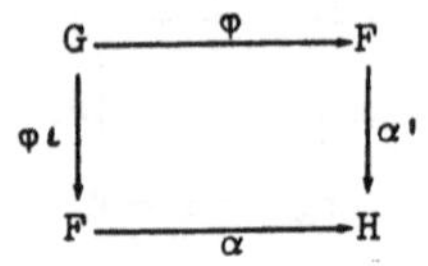

Since $\alpha'\varphi = \alpha\varphi\iota$ and $\iota\iota = 1_G$ we have $\alpha'\varphi\iota = \alpha\varphi$, i.e., the diagram

$$\begin{array}{ccc} G & \xrightarrow{\varphi} & F \\ {\scriptstyle \varphi\iota}\downarrow & & \downarrow{\scriptstyle \alpha} \\ F & \xrightarrow[\alpha']{} & H \end{array}$$

is commutative. Therefore by the definition of pushouts there exists a unique $\iota'\colon H \to H$ such that

$$\iota'\alpha = \alpha' \quad , \quad \iota'\alpha' = \alpha \quad .$$

Therefore $\iota'\iota'\alpha = \iota'\alpha' = \alpha$ and $\iota'\iota'\alpha' = \iota'\alpha = \alpha'$. Therefore by the definition of a pushout we have $\iota'\iota' = 1_H$. Further

$$\iota'\psi = \iota'\alpha'\varphi = \alpha\varphi = \alpha'\varphi\iota = \psi\iota \quad .$$

Thus ψ is negligible as required.

In many cases the category A has the property that if in a pushout diagram φ_1 is a monomorphism then so is α_2. Therefore if both φ_1 and φ_2 are monomorphisms then so is their sum $\alpha_1\varphi_1 = \alpha_2\varphi_2$. It follows that those elements of Φ/Ψ which are represented by monomorphisms $\varphi\colon G \to F$ form a subgroup of Φ/Ψ. This latter subgroup is precisely that described in Chapter VIII, §3, and §4.

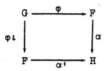

REFERENCES

[1] J. W. Alexander, Note on Riemann Spaces, Bull. Am. Math. Soc., 26 (1920).

[2] J. W. Alexander, A Matrix Knot Invariant, Proc. Nat. Acad. Sci. U.S.A., 19 (1933), pp. 272-275.

[3] E. Artin, Theorie der Zöpfe, Abh. Math. Sem. Univ. Hamburg, 4 (1925), pp. 47-72.

[4] G. Birkhoff and S. MacLane, A Survey of Modern Algebra, MacMillan Co., 1953.

[5] Blanchfield, Senior Thesis, Princeton University, 1949.

[6] H. Cartan and S. Eilenberg, Homological Algebra, Princeton University Press, 1956.

[7] P. E. Conner, On the Action of a Finite Group on $S^n \times S^n$, Ann. of Math., 66 (1957), pp. 586-588.

[8] R. H. Crowell, Genus of Alternating Link Types, Ann. of Math., 69 (1959), pp. 258-275.

[9] R. Crowell, The Group G'/G'' of a Knot Group G, Duke Math. J., 30 (1963), pp. 349-354.

[10] R. Crowell, The Annihilator of a Knot Module, Proc. A.M.S., 15 (1960), pp. 696-700.

[11] R. Crowell, Corresponding Group and Module Sequences, Nagoya Math. J., 19 (1961), pp. 27-40.

[12] R. Crowell, and R. Fox, An Introduction to Knot Theory, Ginn and Co., 1963.

[13] M. Dehn, Über die Topologie des drei-dimensionalen Raumes, Math. Ann., 69 (1910), pp. 137-168.

[14] M. Dehn, Die beiden Kleeblattschlingen, Math. Ann., 75 (1914), pp. 402-413.

[15] R. H. Fox, On the complementary domains of a certain pair of inequivalent knots, Ned. Akad. Wetensch. Indag. Math., 14 (1952), pp. 37-40.

[16] R. H. Fox, Knots and Periodic Transformations, Topology of 3-Manifolds and Related Topics, Prentice-Hall, 1961.

[17] R. H. Fox, Some Problems in Knot Theory, Topology of 3-Manifolds and Related Topics, Prentice-Hall, 1961.

[18] R. H. Fox, A Quick Trip Through Knot Theory, Topology of 3-Manifolds and Related Topics, Prentice-Hall, 1962.

[19] R. H. Fox, Free differential calculus I. Derivation in the free group ring, Ann. of Math., 57 (1953), pp. 547-560.

[20] R. H. Fox, Free differential calculus II. The isomorphism problem, Ann. of Math., 59 (1954), pp. 196-210.

[21] R. H. Fox, Free differential calculus III, Subgroups, Ann. of Math., 64 (1956), pp. 407-419.

[22] R. H. Fox, The homology characters of the cyclic coverings of the knots of genus one, Ann. of Math., 71 (1960), pp. 187-196.

[23] R. Fox and J. Milnor, Singularities of 2-Spheres in 4-Space and Equivalence of Knots, Bull. A.M.S., 63 (1957), p. 406.

[24] H. Gluck, The Reducibility of Imbedding Problems, The Topology of 3-Manifolds and Related Topics, Prentice-Hall, 1961.

[25] H. Freudenthal, Ueber die Enden topologischer Raume und Gruppen, Math. Zeit., 33 (1931), pp. 692-713.

[26] C. H. Giffen, Princeton Ph.D. Thesis, 1964.

[27] A. Haefliger, Knotted (4k-1)-Spheres in 6k-Space, Ann. of Math., vol. 75, no. 3 (1962).

[28] M. Hirsch and L. Neuwirth, On Piecewise Regular n-Knots, Ann. of Math. (to appear).

[29] H. Hopf, Enden offener Raume und unendliche diskontinuierliche Gruppen, Comment. Math. Helv., 16 (1944), pp. 81-100.

[30] K. Iwasawa, Einige Sätze Über freie Gruppen, Proc. Japan Acad., 19 (1943), pp. 272-274.

[31] S. Kinoshita, Alexander polynomials as Isotopy Invariants II, Osaka Math. J., 11 (1959), pp. 91-94.

[32] S. Kinoshita, On the Alexander Polynomial of 2-Spheres in a 4-Sphere, Ann of Math., vol. 74, no. 3 (1961), pp. 518-531.

[33] K. A. Kurosh, The Theory of Groups, vol. I, II, Chelsea, New York, 1955.

[34] W. B. R. Lickorish, A Representation of Orientable Combinatorial 3-Manifolds, Ann. of Math., vol. 76 no. 3 (1962).

[35] R. Lyndon, Cohomology Theory of Groups with a Single Defining Relation, Ann. of Math., 52 (1950), pp. 650-665.

[36] W. Magnus, Untersuchungen über einige unendliche diskontinuierliche Gruppen, Math. Ann., 105 (1931), pp. 52-74.

[37] J. Milnor, A Duality Theorem for Reidemeister Torsion, Ann. of Math., 76, no. 1 (1962).

[38] J. Milnor, On the total curvature of knots, Ann. of Math., 52 (1950), pp. 248-257.

[39] E. E. Moise, Affine Structures in 3-Manifolds V. The triangulation theorem and Hauptvermutung, Ann. of Math., 56 (1952), pp. 96-114.

[40] K. Murasugi, On the genus of the Alternating knot I, II, J. Math. Soc., Japan, 10 (1958), pp. 94-105.

[41] K. Murasugi, On a Certain Subgroup of the Group of an Alternating Link, Am. Journal of Math., vol. 85, no. 4 (1963).

[42] K. Murasugi, On the Definition of the Knot Matrix, Proc. of the Japan Academy, vol. 37, no. 4 (1961).

[43] D. R. McMillan Jr., Homeomorphisms on a Solid Torus, Proc. A.M.S., vol. 14, no. 3 (1963), pp. 386-390.

[44] L. Neuwirth, Interpolating Manifolds for Knots in S^3, Topology, vol. 2 (1964), pp. 359-365.

[45] L. Neuwirth, The Algebraic Determination of the Genus of Knots, Am. J. of Math., vol. 82, no. 4 (1960), pp. 791-798.

[46] L. Neuwirth, An Alternative Proof of a Theorem of Iwasawa on Free Groups, Proc. Cambridge Phil. Soc., vol. 57, Part 4 (1961), pp. 895-896.

[47] L. Neuwirth, The Algebraic Determination of the Topological Type of the complement of a Knot, Proc. A.M.S., vol. 12, no. 6 (1961), pp. 904-906.

[48] J. Nielsen, Om Regnig med ikke-kommutative Faktorer og dens Anvendelse i Gruppeteorien, Mat. Tidsskrift B (1921), pp. 77-94.

[49] J. Nielsen, Untersuchungen zur Topologie der geschlossen zweiseiteigen Flächen I, Acta. Math., 50 (1927), p. 266 satz 11.

[50] C. D. Papakyriakopoulos, On solid tori, Proc. London Math. Soc. (3), 7 (1957), pp. 281-299.

[51] C. D. Papakyriakopoulos, On Dehn's Lemma and the asphericity of knots, Proc. Nat. Acad. Sci. U.S.A., 43 (1957), pp. 169-172.

[52] C. D. Papakyriakopoulos, On the ends of knot groups, Ann. of Math., 62 (1955), pp. 293-299.

[53] E. S. Rapaport, On the commutator subgroup of a knot group, Ann. of Math., 71 (1960), pp. 157-162.

[54] K. Reidemeister, Knotentheorie, Erg. d. Math., 1 (1923), no. 1, reprint, Chelsea, New York, 1948.

[55] O. Schreier, Über die Gruppen $A^aB^b = 1$, Abh. Math. Sem. Univ. Hamburg, 3 (1923), pp. 167-169.

[56] H. Schubert, Die eindeutige Zerlegbarkeit eines Knotens in Primknoten, S.-B. Heidelberger Akad. Wiss. Math. Nat. kl. 3 (1949), pp. 57-104.

[57] H. Seifert, Über das Geschlecht von Knoten, Math. Ann., 110 (1934), pp. 571-592.

[58] H. Seifert and W. Threlfall, Lehrbuch der Topologie, Leipzig und Berlin, Teubner, 1934.

[59] J. Stallings, On Fibring Certain 3-Manifolds, Topology of 3-Manifolds and Related Topics, Prentice-Hall, 1961.

[60] J. Stallings, Unpublished.

[61] H. Tietze, Über die Topologische Invarianten mehrdimensionalen Mannigfaltigkeiten, Monatshefte für Math. u. Physik, 19 (1908), pp. 1-18.

[62] G. Torres and R. H. Fox, Dual presentations of the group of a knot, Ann. of Math., 59 (1954), pp. 211-218.

[63] H. Trotter, Homology of Group Systems with Applications to Knot Theory, Ann. of Math., vol. 76, no. 3 (1962).

[64] H. Trotter, Non-invertible Knots Exist, Topology, vol. 2 (1964), pp. 275-280.

[65] H. F. Trotter, Periodic Automorphisms of Groups and Knots, Duke Math. J., vol. 28 (1961).

[66] E. R. van Kampen, On the connection between the fundamental groups of some related spaces, Am. J. Math., 55 (1933), pp. 261-267.

[67] A. H. Wallace, Modifications and Co bounding Manifolds, Can. Jour. Math., 12 (1960), pp. 503-528.

[68] J. H. C. Whitehead, Simplicial Spaces, Nuclei and m-groups, Proc. of the London Math. Soc., 45 (1939), pp. 243-327.

[69] J. H. C. Whitehead, On the asphericity of regions in a 3-sphere, Fund. Math., 32 (1939), pp. 149-166.

[70] J. H. C. Whitehead, Combinatorial Homotopy I, Bull. A.M.S., 55, no. 3 (1949), pp. 213-245.

[71] H. Zassenhaus, Group Theory, 2nd Edition, Chelsea, New York, 1949.

[72] E. C. Zeeman, Notes on Combinatorial Topology, 1963.

[73] E. C. Zeeman, A Piecewise Linear Map is Locally a Product (to appear).

[74] H. Zieschang, On a Problem of Neuwirth Concerning Knot Groups, Doklady, Acad. of Sci. USSR, Translation, vol. 4, no. 6, pp. 1781-1783.

INDEX

Lightning Source UK Ltd.
Milton Keynes UK
UKHW041212160320
360413UK00001B/125

9 780691 079912